WAR IN THE EIGHTIES

WAR IN THE EIGHTIES

Edited by BRIAN MACDONALD

WAR IN THE EIGHTIES: MEN AGAINST HIGH TECH

THE CANADIAN INSTITUTE OF STRATEGIC STUDIES

Suite 100, 175 Bloor Street East
Toronto, Canada M4W 1E1

The Canadian Institute of Strategic Studies seeks to satisfy a need for a body of informed opinion on strategic and national security issues and to stimulate greater public awareness of the strategic significance of national and international developments. It provides a public forum for discussions to enhance the knowledge of the Canadian public on strategic matters, and through educational and informational activities it seeks to improve the basis for informed choice and decisions by the public of Canada and its leaders on strategic questions and national security matters. The Institute maintains an independent posture and is not an advocate of any particular interest.

ISBN 0-919769-13-6

CANADIAN CATALOGUING IN PUBLICATION DATA

Main entry under title:
War in the eighties

Proceedings of a conference held in Toronto, Ont.,
Nov. 19-20, 1982.
ISBN 0-919769-13-6

1. Military art and science – Congresses. 2. War – Congresses. 3. Technology – Congresses.
I. MacDonald, Brian, 1939- II. Canadian Institute of Strategic Studies.

U104.W37 355'.02 C83-098352-X

Produced by Peg McKelvey
Type by Arjay
Printed and bound in Canada

CONTENTS

BRIAN MACDONALD

INTRODUCTORY REMARKS

Ladies and gentlemen, it is my great pleasure on behalf of the Canadian Institute of Strategic Studies, to welcome you to this our twelfth seminar devoted to the examination of the strategic issues which are of significance and importance to Canada. We are encouraged greatly by the increased attendance and by the fact that some 25 per cent of those attending this seminar were not members of the Institute at the time of last spring's seminar. Such a growth in interest in the work of the Institute is extremely heartening.

The Institute, as you know, is the only organization in Canada devoted solely to the study and discussion of all matters relating to the national security and strategic interests of Canada, and with the aim of helping to create a better informed and attentive public in Canada with respect to these matters. To this end our seminars are public, the media are present and welcome, and the proceedings are recorded and subsequently published. I would ask you, therefore, if during the question periods and the open forums you choose to contribute a question or a comment to clearly identify yourself so that you may be properly recorded in the proceedings.

It is useful in our role as an Institute of Strategic Studies to from time to time review the meaning of the term strategy. In our last seminar, for example, Bernard Thillaye spent some time in a cogent analysis of the term from a perspective drawn from the military strategic paradigm. It is, perhaps, useful to observe that there is another strategic paradigm which has been created in the various schools of business administration and that strategists of both may gain useful insights from the other.

Two theoreticians of the business paradigm, Charles Hofer and Dan Schendel, argue that 'the basic characteristic of the match which an organization achieves with its environment is called its strategy.' 'Strategy is the match between an organization's resources and skills and the environmental opportunities and risks it faces and the purposes it wishes to achieve.' Strategists are, therefore, compelled to interpret the events of the future and to fashion policies which will allow their organizations to cope with that future.

At our last seminar we had the opportunity to examine not only the process by which defence policy has been developed in Canada, but to examine analyses of that policy and its strengths and weaknesses by Parliamentary and Senate committees. Insofar as policies may be considered as the manifestation of a national strategy we were also assessing the defence strategy of the Government of Canada. Meanwhile, in the war between Argentina and Great Britain, a more violent and costly analysis of the defence strategies which had been adopted by those two nations was taking place. The results of that war clearly indicated the success of the match which each nation had achieved with respect to the threats and opportunities that lay in their future environments. As we will discover this morning, each had successes and each had failures in coping with that future environment.

Brigadier Millen, the British Defence Advisor to Canada, suggested in Ottawa at the beginning of the National Speaker's Tour that the paramount lesson to be learned from the Falklands war is that the future will be unexpected. The Greek philosopher Herakleitos of Ephesos, writing in 500 BC, said 'You cannot step twice into the same river; for fresh waters are forever flowing in upon you. The sun is new every day.' Theirs, however, is not the counsel of despair but an exhortation to prepare for that uncertain future so that the unexpected may be coped with. As that noted Canadian, Sir William Osler, pointed out: 'When schemes are laid in advance, it is surprising how often the circumstances fit in with them.' Such, then, is the role of the strategist.

This seminar continues that process of strategic reasoning which we began in the last. In that seminar we examined Canada's existing defence strategies. In this seminar we move to a consideration of the future environment with which we must achieve a match. Our focus is placed upon the technological aspects of that environment, and our nation's technological resources. We will also consider man as an additional resource, as well as the interface which must exist between man and machine, if we are to achieve that match between our nation and its future environment which is required of a successful strategy.

We must realize that our analysis of our environment will be limited, even in the technological sense. We will not, for example, deal with the developing technology of defence on the space frontier which has profound implications for the strategic balance which has hitherto existed. Nor will our seminar deal with the political and ideological components of our future environment. Nor, for that matter, will it deal with the economic aspects of that environment. That examination we will leave for future seminars of the Institute.

Ladies and gentlemen, it is time to begin the action of the seminar.

S. ROBERT ELLIOT

A STRATEGIC ASSESSMENT OF THE FALKLANDS WAR

LCol Brian MacDonald

Our first speaker has arrived in Toronto as the last step in his national speaker's tour - a tour that has performed at least as well as we had hoped, and much better than we had expected. He has been heard by over fourteen hundred Canadians in eight cities across our nation. With this morning's attendance of over a hundred, he will have been heard by some fifteen hundred people.

Major Elliot is doubly welcome. He represents the prestigious International Institute for Strategic Studies as its Information Research Officer and as a senior policy researcher in that institute, and is also welcome as a former Canadian officer who ended his career as an intelligence officer on an exchange posting in Britain. He has had a very large number of articles published in a variety of journals. He is responsible for the publication, *The Military Balance*, of the International Institute for Strategic Studies.

Major S. R. Elliot

Thank you very much, Brian.

When I was given this task, the general parameters of the speech were to be a quick review of the strategy of the Falklands Islands war, then the war itself, the lessons learned by Britain and the application of those lessons to Canada. That is the format I intend to follow.

The Argentine junta's strategy required it to take over a territory it long regarded as its own. The islands were within easy sailing distance and just within the range of limited air cover from the bases in south Argentina. The junta was sure it could count on Third World support, and benevolent or neutral Soviet and United States attitudes. In the opinion of the military chiefs, Britain had been seen as being unwilling to support the colony. The distance, and British defence cuts, made a military reaction unlikely. A swift coup de main would force Britain into meaningful negotiations towards a transfer of the islands to Argentine political as well as de facto control.

The British attitude towards the Falklands was indeed ambivalent. There was by no means universal belief in the validity of the British claim to sovereignty. The islands themselves represented a minor, and relatively unproductive vestige of empire, which could neither be defended nor made self-sufficient. British defence doctrine called for priority to be given to operations within NATO, and minimal support for operations outside the NATO area. The government misread the pressures on the Argentine leadership – social, economic and political – which led to a foreign adventure being the only practical avenue by which to gain and keep popular support. Having mismanaged the initial crisis, the British government was faced with a dilemma. Acceding to a political and military defeat would have been totally unacceptable to a nation resentful of aggression against its own people, and so would lead to the fall of the Thatcher government. The alternative, to fight a war to recover a piece of unwanted territory, would be expensive and difficult, and victory was by no means certain. Operational considerations did not favour the British defence forces. There were no air or naval bases within easy reach. The nearest – Ascension, 4000 miles from the Falklands – had only a limited air facility. It offered an anchorage only, not a port. A heavy casualty toll would also put the government at risk. The basic dispute at issue, that is, sovereignty, would remain unsolved.

The Falklands crisis began on or about 27 February 1982 with yet another inconclusive discussion between British and Argentine representatives over the islands' future. The junta reviewed its long-standing contingency plans for occupying the territory. The navy took a leading part in this review. The landing, on March 19, of a party of metal salvage workers on South Georgia, and their raising of the Argentine flag, was a navy-inspired gesture intended as a provocation to Britain and a challenge to moderates within the junta.

As a precaution, Prime Minister Margaret Thatcher, on March 29, ordered a nuclear fleet submarine to the South Atlantic. The handover of the Royal Marine garrison in the islands was delayed. The British First Sea Lord, Sir Henry Leech, on his own initiative, set planning in train to send a task force as a counterstroke. The only field force available was 3 Commando Brigade, Royal Marines, comprising 40, 42 and 45 Commandos. RAF aircraft were sent to Ascension with airfield equipment and stores on April 1.

Argentine marines landed near Port Stanley by helicopter about 0430 on April 2. The garrison surrendered at 1000 and the governor was on a plane for Britain via Argentina by 1800. Argentine marines arrived at Gritviken, South Georgia, the next day. The 21 British marines shot down two helicopters and badly damaged a corvette with their anti-tank rockets before surrendering.

On April 3, the British House of Commons agreed that the islands would be recovered. The United Nations unanimously approved resolution 502 which called upon Argentina to withdraw and for both sides to enter into negotiations on the future of the Falklands. Two days later the carriers *Hermes, Invincible,* and the royal fleet auxiliary tanker *Pearleaf* sailed. The Ministry of Defence had begun to requisition and to charter an additional fleet of civilian vessels. The commander, 3 Commando Brigade, Brigadier Julian Thompson, was given in addition two parachute battalions each 5 Brigade. The force left on April 6, 4820 strong. There followed unproductive diplomatic initiatives by United States Secretary of State Haig, the Peruvian president and the UN Secretary General. British diplomacy rallied United States popular support and convinced the EEC to ban arms exports to, and commercial imports from, Argentina. Britain imposed a military exclusion zone around the Falklands on April 7, the implication being that a nuclear submarine had arrived. It raised it to a total exclusion zone on April 28 and on May 7, imposed a blockade on Argentine coastal waters.

The Falklands operation, code named 'Corporate,' was controlled by Admiral Sir John Fieldhouse, C-in-C Fleet, from his headquarters at Northwood, near London. The task force commander was Rear Admiral R.J. 'Sandy' Woodward. The ground operation was under the command first, of Brigadier Thompson, later, of Major General Jeremy Moore, Royal Marines.

The recapture of South Georgia was the first priority. It would provide a base well out of range of Argentine air, 800 miles east of the Falklands, with a good sheltered anchorage and some shore installations. Its political status was clear: it is not part of the Falklands. Its recapture would be good for British morale and correspondingly bad for Argentina.

The attacking force was M Company, 42 Commando - 110 men - led by the Commando second-in-command, Major Sheridan, just back from an Arctic warfare exercise, and some 70 special boat service (SBS) and special air service (SAS). The force was flown to Ascension and loaded into the county class destroyer *Antrim*, the antarctic patrol vessel *Endurance*, and the RFA tanker *Tidespring*. The SAS and SBS were to reconnoitre the island at Gritviken and Leith, moving via Husvik and Stromness.

On April 21, two Wessex helicopters were sent out with three four-man SAS patrols. The first sortie failed but the second landed on Fortuna glacier. The weather closed in. Two attempts were made the next day to lift them off. Both Wessex helicopters crashed. On the third attempt, *Antrim*'s Wessex picked up the 13 SAS, plus the two crews, and brought all 17 back to *Antrim* in a spectacular piece of flying that won the pilot, Lieut Cdr Ian Stanley, the DSC. (The capacity of a Wessex is normally five.) Two SBS squads from *Antrim*, which had landed in Hound Bay to move to Gritviken via Moraine Fjord, had also to be rescued. A third SAS attempt on 23 April with five Gemini assault boats came near to disaster when two of their notoriously unreliable Johnson outboard motors failed. One patrol was picked up by *Antrim*'s helicopter. The other landed on the last point of land before Antarctica and walked back to Leith. An obvious lesson: if the equipment is unreliable in peacetime it will surely fail in war.

These setbacks, plus reports that an Argentine submarine was in the area, worried Admirals Fieldhouse and Woodward, but *Antrim*'s skipper by then had gone over to the offensive. At 0630 on April 25, the submarine *Santa Fé* was spotted, incoming to Gritviken, on the surface. It was depth-charged and driven ashore. Sheridan managed to refocus naval attention on his mission, and although his main force was still some hours away, he cobbled together a 75-man scratch force of his marines and the SAS and SBS, and went ashore supported by gunfire from *Antrim* and the frigate *Plymouth*. The Argentine garrison at Gritviken ran up white flags as he came ashore. Leith surrendered the next day and the 137 prisoners were taken aboard *Tidespring*. There had been two Argentine casualties, one sailor wounded in the attack on the *Santa Fé*, and another who was shot owing to a misunderstanding. There were no British casualties. This was an Argentine mistake. It gave the British a cheap victory.

The British landings on the Falklands proper began on May 1 with the deployment of SAS and SBS patrols to reconnoitre landing areas and monitor Argentine activity. That morning also, a single Vulcan bombed Port Stanley airfield. Harriers followed during the afternoon and also hit Goose Green. On May 2, the nuclear-attack submarine *Conqueror* sank the cruiser *General Belgrano* with two Mk 8 torpedoes. *Belgrano* sank in about an hour. The torpedo strike, and losses from an overturned liferaft later, killed 368. That afternoon, also, Harriers sank one Argentine patrol boat and damaged another.

On May 4 *Sheffield*, which had been deployed as a radar picket west of the islands, was hit by an air-launched Exocet and abandoned after attempts to quell the resultant blaze had failed. Twenty were killed; 24 hurt. On May 9 the fleet captured an intelligence collector, the *Narwal*. On May 11, the frigate *Alacrity* sank a transport in Falkland Sound.

An SAS raid on May 14, supported by gunfire from *Glamorgan*, destroyed 11 aircraft on Pebble Island. Two more Argentine marine transports were caught on May 16; one was sunk, the other driven ashore.

The landings began in San Carlos water on May 21: 40 Commando and 2 Para went to San Carlos and Sussex Mountain; 45 Commando and 3 Para went to Ajax Bay and Port San Carlos; 42 Commando and the artillery came ashore at Ajax as followup. The day was clear, with no cloud to give cover to the landing. About noon the Argentine air force began to counterattack. Some 80 sorties arrived over the Sound that afternoon. The Type-21 frigate *Ardent* was badly hit, losing 22 killed and 30 injured. Five ships were damaged. Fourteen Argentine aircraft and two helicopters were shot down.

On May 23, Argentine air hit the frigate *Antelope*. The bomb, which landed in her engine room, did not then explode but did so while the expert tried to defuse it. Fortunately casualties were light. Eight aircraft were shot down.

On May 25, Argentina's national day, the destroyer *Coventry* was sunk with the loss of 21 killed and 21 injured. *Atlantic Conveyor* was hit by an Exocet diverted from *Invincible*. All her load of harriers had been flown off but three of the four Chinook and all six of the Wessex transport helicopters she carried were still aboard, as were the 20 tons of tentage, engineer vehicles and material for the Harrier landing pads. Argentina lost 10 aircraft and this attrition finally proved too much. Thereafter only small raids took place. One, unfortunately, was on the brigade maintenance area, where it destroyed a considerable quantity of mortar and other ammunition.

Brigadier Thompson had planned to send his forces against Port Stanley in a two-pronged attack along both the northern and southern routes around the island. For this, he needed the helicopters to give his plan flexibility and support. He wanted 5 Brigade but before it arrived he was ordered forward.

The southern operation started with 2 Para's capture of Port Darwin and Goose Green. The Argentine regiments occupied a ridge running obliquely along the narrow isthmus connecting East Falkland with Lafonia. Approaches to it were flat, with virtually no cover. Defensive fire was well planned and co-ordinated in depth. The Paras had three 105mm field guns plus fire from the frigate *Arrow*'s single 4.5 inch gun in support. Harrier support had been planned, but the low ceiling prevented it from being used until late in the afternoon.

Despite early difficulties and the loss of the Battalion commander LCol H. Jones, by the end of the day the 450 men of 2 Para had defeated some 1600 Argentines, perhaps 1300 of them combatants. It cost them 17 killed, 34 wounded. Argentine fatal casualties were 50, mostly from the cluster bombs from the Harriers. The three field guns fired some 900 rounds; *Arrow* 235.

The initial advance on the northern flank also began on May 28. 45 Commando was ordered to Douglas; 3 Para to Teal Inlet. The 45 Commando "yomp" (that's their word for a pleasant walk in the country) started with the men carrying up to 120 pounds of gear. It took them, eventually, some 60 miles to the mountain Two Sisters. The Paras took 24 hours non-stop to get to Teal, about one-third of the time Brigade HQ had thought necessary. Major General Moore took command of the land battle on May 30.

On May 31, K Company began 42 Commando's lift by helicopter to Mount Kent. It occupied Kent and Mount Challenger by 3 June. Those two features had been abandoned by the Argentines to back up Goose Green and Darwin: again, a tactical error. Headquarters 3 Brigade moved to Kent by May 7. The brigade was given three batteries, 7, 9, and 79 of 29 Commando Regiment, RA – 18 105mm guns – with 1000 rounds per gun, all moved up by helicopter.

5 Infantry Brigade began landing at San Carlos on 1 June. By this time this brigade was something of a scratch force. Its two para battalions had gone to 3 Brigade, replaced by 2nd Battalion Scots Guards and 1st Battalion Welsh Guards. The Guards had been on public duties – stamping up and down in front of Buckingham Palace, and there is some considerable doubt as to their standard of training and fitness. Its third Battalion, of course, was 1/7 Gurkhas. One gets the impression that this brigade had originally been intended solely as a garrison and was put into combat only after the size of the Argentine garrison became known and after the losses from *Atlantic Conveyor*. The initial estimate of that Argentine garrison had been 3000. By this time, estimates had been raised to between 12 000 and 16 000.

When Brigadier Wilson, Commander 5 Brigade, heard that the Argentines had withdrawn from Bluff Cove and Fitzroy Harbour, he sent two companies of 2 Para to take over. He then tried to march the guards forward but they were halted. The helicopters were lifting supplies forward for 3 Brigade and PWs back to the concentration areas and were not available to support him. The only alternative was a sea lift. Owing to battle damage, there were insufficient escorts to protect this lift but the gamble was considered acceptable, providing most of the move was at night.

The Scots Guards were to be moved in *Intrepid* to Lively Island. From Lively they were to be lifted to Bluff Cove in four utility landing craft. It took them five hours due to rough water, winds gusting to 70 knots, and navigation difficulties. In order to cut down the time spent in open boats *Fearless* took the Welsh to Direction Island, but the weather was so foul that the LCU could not come out to meet her. She loaded as many as possible into the two LCUs she carried, slipped them, and returned to San Carlos with the rest of the battalion.

On June 7, the Welsh were trans-shipped to *Sir Galahad*, together with a number of other brigade unit personnel. It was daylight before she left San Carlos. She found the Bluff Cove channel too narrow and diverted into Fitzroy, where *Sir Tristran* was already unloading. But there were not enough landing craft for both tasks. Further, from Fitzroy the troops would have had to walk the 12 miles around the inlet to Bluff. They elected to wait until the LCUs were free. The Skyhawks and Mirages arrived first. Sixty-three were killed, 46 injured. The navy admitted fault for not giving protection. The Guards were criticized for not going ashore and for doing nothing to protect themselves. There were other reasons. The Harrier cover had been withdrawn, but nobody had been told. An air raid imminent warning had not reached the ships. The Rapier had not been set up on shore early enough, although it did come into action, claiming one killed.

On the northern flank, 3 Brigade led the assault with five artillery batteries and some naval guns in support. 3 Para captured Mount Longdon in a night attack on June 12, losing 23 killed and 45 wounded. One of its NCOs, Sgt McKay, won a VC taking out a machine-gun nest. In taking Two Sisters, 45 Commando lost four killed and eight wounded. Mount Harriet fell to 42 Commando with fire support from two companies of 40 Commando and what was left of the Welsh Guards. They used their Milan anti-tank missiles against machine-gun posts and an outflanking movement around the hill to attack the defenders from the rear. They lost one killed, 13 wounded and took 250 prisoners, many of them wounded. In 5 Brigade's follow-up assault, the Gurkhas took Mount William which, to their intense disgust, was largely empty. On June 13, 2 Para took Wireless Ridge with support from the four Scimitars and two Scorpions of the troop of Blues and Royals. Later that night, the Scots Guards took Mount Tumbledown, held by an Argentine marine battalion which literally had to be dug out with the bayonet. The engagement lasted some 14 hours. The Scots lost nine killed, 41 wounded; the Argentines 30 killed and 40 prisoners, many of them wounded. The Welsh finished their war by taking a deserted Sapper Hill. Many casualties in these last actions were caused by the little plastic mines that had been seeded indiscriminately on the approaches, often by helicopter, and which are still there.

At 2100 June 14, General Menendez surrendered. Britain lost 255 killed, 18 of whom were civilian crews aboard the royal fleet auxiliaries and *Atlantic Conveyor*. The number of Argentine casualties is still unknown.

It has been said that this war was won not so much by superior equipment as by professionalism. As an example, the decision of the First Sea Lord to anticipate his Prime Minister's requirements and to initiate planning for the task force is in the highest professional tradition of the RN. And this is reinforced by the fact that the first units were at sea within three days of the decision to send them. The navy mobilized 23 warships and 26 fleet auxiliaries, and some 54 merchant ships were either chartered or requisitioned. Many of these latter vessels had to be modified. Dockyard workers, some of them with dismissal notices in their pockets because their dockyards were scheduled to close, worked round the clock to get the force to sea. Some modifications were unscheduled. The master of the ro-ro ferry *Elk*, and his RN liaison officer, for example, refused to go to war unarmed, and by scrounging material and some engineers constructed gun tubs, mounting in them some Bofors 40mm guns. He had to be dissuaded from putting in a ski jump for a Harrier.

Moving a large naval force 8000 miles requires extensive logistic support. Refuelling required naval fittings to be mounted in *Canberra* and the *Queen Elizabeth II*, among others. Transfers of personnel and stores at sea became almost routine. That there were no major accidents reported despite some very severe sea conditions is a mark of the standard of seamanship that prevailed. Gaining that seamanship is difficult in a small navy without putting an unacceptable strain on the crew members and their families. In wartime, this can be accepted; in peacetime it cannot. Canada does not have enough sailors for the ships it now has.

Each ship must be able to protect itself through a wide spectrum of air attack. Because the concept of war in the NATO environment has stressed specialization, with ships acting as part of a team, they become vulnerable if, for any reason, the team is incomplete. In the Falklands the enemy used mass attacks to swamp the defences. Even though his aircraft were obsolete, they did have some successes. Combat pilots will tell you that, no matter how good the defences, something will invariably get through, and so it proved, albeit at high cost.

Fleet operations in the North Atlantic are becoming more dangerous, not only from the increasing number of Soviet aircraft assigned to long-range fleet protection, but also as a result of the long-term Soviet policy of mounting a wide range of surface-to-surface missiles in its combat vessels. Only four of Canada's DDHs have even a simple surface-to-air missile – the Sea Sparrow, a design which first emerged in 1964. Is this weapon still able to do its job? Are the electronics which support it adequate? The Falklands proved that there has to be a layered defence, with its various components so interlocked that, even if one element fails, there is always something to back it up. This is not possible in today's Canadian maritime force.

An interesting facet of the Falklands war was the re-emergence of the naval gun as a weapon of war. Not only were these used against shore targets, on their own and in support of the military, but also against minor naval targets. *Glamorgan*, for example, put some 12 000 rounds through her twin 4.5 inch guns.

She was supporting 45 Commando's attack on Two Sisters when she was hit by a land-based Exocet. The frigate *Arrow* was supporting 2 Para at Goose Green. The feeder jam which took her out of action shortly after the battle started may indirectly have contributed to the check in that unit's initial assault. Most ships in the force had but one 4.5; the Broadsword class had missiles only. All amphibious assaults need fire support so long as the troops are in range of the naval guns. Earlier this year I observed the paradox that the best ship for the support role was on the other side – the *General Belgrano*. I was sharply told that the navy had decided to eliminate the gun some years ago. I submit that there is still a need for a gun in a number of situations; for example, where a missile would be more weapon than the task requires; where electronic countermeasures could nullify the missile attack; where there are considerations of expense, such as a shot into a small menace to navigation.

As a possible scenario, consider the naval support of a Canadian force ashore in Norway, under the terms of the CAST commitment. I am not sure how such a force is expected to move and fight, once landed. The Norwegian terrain would suggest that much of the action could be relatively near the coast, and the ground forces never have enough artillery support. The questions arise: are the present Canadian fleet units capable of performing this function? Do their guns have the range? Do they ever practice it? Is there provision for the control and co-ordination of such fire? New vessels are being considered for construction. What is the weapons mix for them? Is it adequate?

A feature of the war which came as a shock, particularly to civilians, was the damage that results from both missiles and iron bombs. The combination of kinetic energy and high explosive plus burning fuel either from the tanks of the target vessels or from the missile itself is devastating. No practical amount of armour will protect against present missiles. No ship can be made unsinkable, but it should be possible to improve on current construction and ensure the maximum protection possible.

There are three levels of marine anti-air defence. The first layer is the weapon system, including electronics, that either nullifies or destroys the enemy weapons platform, or the weapons deployed from it. The next defensive layer is the ship herself and this is a function of design. The third layer comprises the damage-control measures intended to limit the effect of a hit. Some of these may be active, including warning devices, and trained fire-control teams. Others are passive, measures intended to starve a fire in the vessel.

The Exocet which hit *Sheffield* destroyed the control centre, the communications systems, and the power supply to her pumps. Damage was compounded by the effects of fire on the ship's structure and on such fittings as mattresses and the so-called non-inflammable wooden panelling, all of which produced clouds of poisonous smoke. Even those fighting the fire were at risk. Smoke hoods were inadequate and men used their CW respirators. Modern synthetic clothing sticks to the skin when it ignites. The burns that result are quite horrible. The old cotton and woolen garments could at least be stripped off.

Fire-fighting methods seem not to have changed since the wooden walls of Nelson's navy. A fuel fire, as any fireman will tell you, spreads when you pump water on it. In a picture of *Sheffield* burning there is a raft with a man pumping a little stream of water into the wreck. Was he, with all the best will in the world, merely adding to the problem? Is there perhaps a way of fitting ships with package foam kits which would operate automatically, which would require little checking and which would be able to remain, inert, through a wide range of temperatures? Perhaps more opportunity should be given to those who have to fight the ship to get their views across to the constructors on such matters as weapons fit, fire safety, and electronics.

The Royal Navy had two roles. The first – sea control and support – I have discussed. The other was to provide air cover for the task force. The Argentine naval air and air forces were highly professional. The air force pilots had been trained by the Israelis. They had the dash and bravery to be expected of the society from which most of them came. Those who, in the early days of the campaign, recalled the sinking of the *Repulse* and *Renown* by Japanese fighter bombers off Malaya in 1941 were right to be concerned.

The task force had first to secure air superiority. Though early warning of approaching enemy aircraft was essential, no airborne early warning facility was available. The radar picket *Sheffield* was the alternative. Her loss forced Admiral Woodward to deploy his major units east of the Falklands, just on the edge of Argentine aircraft range. The force used its Harriers in contact air patrol, an uneconomic method of deployment. They were flown off and the ships then moved forward into the zone to retrieve them.

The Argentine air force started the war with only 81 of their 224 combat aircraft in service. Only a few could be refuelled in flight, using the two KC-10 tankers on inventory. They did not have external fuel tanks and though they tried desperately, they could buy none. The naval *Super-Etendard*, which had a buddy system for refuelling, had only one Exocet for each of the five aircraft. The Argentine air force lacked a long-range reconnaissance capability. In the beginning, crews had to search for their targets. When the task force moved into Falklands Sound their task was made easier. The pilots knew where it was. They could approach, hidden in the ground clutter of the hills on West Falkland. Their range was so marginal, however, that they still could spend only a limited time over that target. They could not even use their afterburners. Their only protection was surprise. British radar was not infallible. No aircraft in the fleet had a look-down/shoot-down capability. It would have helped, but it would not have replaced an airborne early warning facility.

The Harriers scored most of their 31 recorded kills with the AIM-9L Sidewinder. Of the 27 launches, 24 hits were scored. Only one completely failed. The other kills were to the 30 mm Aden gun. The infra-red missile is now replacing the gun in air combat, but the ability to engage approaching targets is essential. I presume the new CF-18 will have this new weapon?

The Harrier range gave them 40 minutes of station at a distance from the carrier of 200 nautical miles. Their on-board radar was not intended for intercept, but for ships, and most of their sightings were visual. They flew a total of 1893 sorties, 1500 of them official, 282 at night. The total was 2514 flying hours. Only one per cent of the planned sorties had to be abandoned because of unserviceability. Pilots would do four sorties a day and often this involved up to ten hours a day in the cockpit, flying weather with 100 foot ceilings and one-eighth of a nautical mile visibility. Nine Harriers were lost, four through accident, two to enemy SAM and the other three to gunfire. The total number sent into the Falklands was 42: starting with 12 in *Hermes*, eight in *Invincible*. Eight Sea Harriers and 14 RAF G-3 Harriers flew via Ascension with inflight refuelling some 4000 miles to rendezvous. The RAF pilots had never landed on a carrier before, which caused a local cartoonist to show the crew gluing grass mats on the deck of *Hermes*.

A thesis being advanced in the United States is that a carrier must be the nucleus of any navy whose operational responsibility extends outside the range of its own land-based tactical air. Conditions in the Falklands supported this thesis. One of Canada's NATO roles is long-range anti-submarine support. Will this always be in conjunction with another navy or will our navy ever have to operate on its own? The Harrier owed a great deal of its success to the advantage it had of fighting the enemy on its own terms. The enemy faced constraints imposed by distance and the obsolescence of its aircraft. The RN were superior in sheer professionalism, based on NATO-honed skills. Canada's pilots rank with the best, but are they getting enough air time to maintain and improve those skills at 20 hours a month?

Aircraft range and deployment flexibility are every bit as critical to the battlefield as any other weapon range. It is vitally necessary in peace time to develop to the maximum any facility, or pilot skill, which will increase the range of a combat aircraft at minimum cost to its effective payload.

The only essay into strategic interdiction was a failure. One Vulcan bomber dropped perhaps 21 1000-pound bombs and scored one hit. A second sortie missed the runway altogether. To get the two sorties there took ten tankers each. The runway remained open until 1400 on the day of the surrender. The last C-130 got out one jump ahead of the artillery. Throughout the period, Argentine C-130s brought in 435 tons of supplies and evacuated some 264 wounded.

The anti-submarine Nimrod aircraft did long-range patrols over the supply route and there are suggestions that Victors did at least one reconnaissance along the Argentine coast. The Vulcan raid apparently convinced Buenos Aires that a raid on the mainland was possible. It was, in fact, never seriously considered, nor, contrary to popular press speculation, was there any intent to deliver a nuclear strike. One major modification to the Nimrod was reported. After an encounter with an Argentine 707 reconnaissance aircraft near the task force they were given Sidewinders to protect themselves. There may be a similar requirement for Canada's long-range patrol aircraft.

The missile war is more difficult to assess. Though the British have announced the kill claims, they have not said how many of each were fired. The reasons for missile failure are critical, and hence tightly covered by security. The preliminary claims are: Sea Dart, eight; Sea Cat, six; Sea Wolf, five; Rapier, thirteen; Blowpipe, eight; Stringer, perhaps one.

Some problems with Sea Dart have been revealed. The launcher, mounts two missiles only. Launchers have to be reloaded manually on the open deck. The automatic doors to the magazine failed and had also to be closed manually. There was a mix of radar and command systems on the different ships. Some were too slow to collect, collate and transmit the data needed to engage. Sea Dart has to be stored inert and the mandatory gyro warmup time for reload is two minutes, so you could have a ship unarmed for an extensive period. They are adequate at long range but not for low-level targets. Sea Cat were used as a deterrent, though they downed six. Only two vessels had Sea Wolf: *Brilliant* and *Broadsword*. Its computer initially had difficulty when confronted with more than one target, and when it had difficulty, it zeroed everything. A simple modification by the task force technicians cured this. Line-of-sight engagements found the target obscured by the efflux plume – a problem from the cold Antarctic air. A similar problem could exist in the north. This was cured by modifying the guidance to produce an oblique flight. Sea Wolf is a point defence system and it was not found to be capable of protecting anything other than its own ship. In the air-to-surface role, the Sea Skua, which were fired from the Lynx, were successful against low-profile targets in heavy seas. Even the AS12, old though it may be, did have some success.

Rapier took time to set up and adjust. There were generator failures. It had not been intended to use it in a naval environment and its emissions fouled up fleet electronics including Identification, Friend or Foe. Some 45 missiles were fired: there were 13 hits. The Paras were unhappy with Stinger. Four were fired: the one hit was also claimed by a unit using small arms. The Marines found Blowpipe heavy and vulnerable to decoys. One direct hit on a Puccara is confirmed. Claims of five more, including two Mirage and a Skyhawk, are being negotiated.

Despite the fact that most of the damage to the fleet was caused by iron bombs, there is no doubt that the missile warfare age has arrived. Granted, the Argentines did not have adequate electronic counter-measures and so some caution must be exercised in assessing the true value of the missile defence. Nevertheless, it is clear that a closely integrated missile system is the only real answer against mass hostile air and missiles. Long-range anti-air keeps the bomber at a distance and so can inhibit the use of the ordinary bomb, but may not be so successful against the air-launched cruise missile. Considering the fact that the Canadian forces have almost no significant SAM capability afloat or ashore, I suggest we have a lot of catching up to do.

The land battle was strangely reminiscent of 1917. However, there are a number of questions. The British have long practised amphibious operations, yet there seemed to be gaps in this one. The initial loading was rushed and planners may not even have known the vessels into which the material was to be loaded. Some transshipping at Ascension may have sorted out some of that, but obviously not all of it, and ships remained specialists with the risks that that condition entails. With *Atlantic Conveyor*, for example, went all the force tents. It is essential to load tactically. Rapier, for example, had to be manhandled up three decks before it could get ashore at San Carlos. In a hostile air environment, SAM must be one of the first support weapons ashore.

The force seemed to lose momentum once it landed. Inadequate intelligence gathering may have contributed to this. The recce helicopters and the Harriers were committed to other tasks. There were no drones. Special forces patrols were withdrawn just before the landing. One wonders if perhaps there was a failure at the command level to organize an extra intelligence increment. The brigade establishment is not really designed for independent action. And did General Moore's headquarters really get a grip on the action, coming in (as it were) at half time?

The attack on Darwin was said to have been necessary to avoid leaving a large enemy force behind the advancing column. Could a company have held that sector, using deception, to let the rest of the force go on? Thompson left one of his Commandos at Ajax, much to its commander's disgust. Its purpose, apparently, was to guard against an airborne threat. It is not quite clear how Argentina was to mount this, but with no airborne early warning it was a threat that could not be ignored. Thompson's choice of units seemed odd. 40 Commando was much more mobile than the Guards. It should have gone forward.

Goose Green reminded General Moore that artillery support was essential. Despite the lack of transport, this was made available for the final assault. Some ammunition expenditures may be of interest. In ten days, 7 Battery fired over 3500 rounds; 8 Battery, 5180; 79 Battery, 2700. At the end, they were down to six rounds per gun.

The place of the helicopter on the battlefield was never more clearly demonstrated. They were carefully used by the British behind their own lines. When a mistake was made, as, for example, around Port San Carlos where two Gazelles were lost, their vulnerability was clearly demonstrated. Tactics and training on these aircraft must ensure that realistic and practical doctrines apply. Certainly, in the transport role, the value of the heavy-lift helicopter to provide mobility and flexibility is incalculable.

Despite reports at the time, the Argentines had more and better fighting equipment and the units using it had some success. But, in the main, their conscripts were badly handled, badly supplied and badly led. The British had superior motivation and training.

Some of the British specialist programmes include adventure training. This is perhaps something we could well do in Canada. I can see a very real place for such extramural activities as long-range Arctic expeditions, mountain work, canoeing or other small boat exercises, gliding and parachuting, not just for an afternoon on a weekend, but for extended periods with small groups of men under their own officers and NCOS. The cost would be minimal; the benefit to training and morale incalculable.

Once again, it is clear that the most important component in any land battle is the individual soldier. It is his training, esprit de corps, motivation and discipline that wins wars: their lack that loses them. Discipline will take a man a long way, but, in the final analysis it is what the man himself decides to do that is going to tip the balance between success or failure. This is a lesson we all tend to take for granted. We cannot.

The campaign demonstrated that Britain still has the capacity to mobilize and to deploy a show of force out of the NATO area. The world at large has been shown that there are, indeed, limits to tolerance. At the same time, a common British theme, that this would show the world that aggression does not pay, is perhaps a little overstated. Aggression is very much alive and very well in a number of places. The Soviet Union was shown that 'decadent capitalist' nations can still respond to a military challenge. It demonstrated to NATO that the

United States was aware of where its real interests lay. The cost of Washington's support, in terms of damage to its Latin American policy, must not be underestimated.

What it shows Canada, I suggest, is that we, who have not had to use battle skills since Korea, may have much to learn. We pay lip service to the principle that the basic skills of soldiering, developed and enhanced by training, still are effective. The Falklands war has shown that, even at a low and incomplete level, the business of war has become very much more sophisticated. We must identify such significant weaknesses as may exist in our armed forces, equipment, training and the ability of each component to work smoothly and efficiently together. A country whose military forces are not tailored to its commitments is flirting with defeat should a situation ever arise where it has to use them. Are Canada's armed forces now in this position? Regrettably, I know people, as do you, who will answer 'yes' to this position. A more serious question is: How do we remedy their shortcomings?

LCol MacDonald

We now have a period for questions and observations.

Leonard Bertin

I would like to know, now that the problems in electronic control have been identified, whether these are fundamental design problems? Must we scrap what we have and start from the beginning again?

Elliot

Most of these weapons, Sea Slug, Sea Cat, Sea Dart, are first generation weapons. Sea Wolf is a second generation system, a replacement for Sea Cat. Most are improvements on existing weapons. I think we are going to get here, as we have in every other weapon, generational changes, improvements to the existing technology. The Marconi people tell me that they are working on improved simplified control systems. And there are improvements coming onto the drawing board, if not yet in production.

One area that may repay examination is the Russians' system of mounting all their vertical launch weapons below the deck. If you look at pictures of *Kirov*, for example, all you will see on her decks is a row of hatch covers. We, on the other hand, tend to put everything we have high up in the ship which raises the centre of gravity and provides problems of balance.

Ivan Phelan, Grenadier Guards

You have made no reference to communications equipment. Would you comment on that?

Elliot

Frankly, I know very little about it. It was, as far as I know, standard British equipment. The SAS had their own material which has enormous range. They were talking directly to their headquarters in London. Most of the fleet communication was by satellite, and I think that the task force commander found to his chagrin that he was under the direct control of London - he had no flexibility himself. There was a problem here. One of the stories about *Sheffield* was that she was using her rear link and had her radar switched off when the Exocet hit. That may or may not be true. Also, there were three fleets at sea: one a combat fleet, one the amphibious fleet, and the third the support fleet. Communication between all three did present problems.

As far as the infantry was concerned, there was the normal battalion rear link in use. I don't think there were any problems with it.

David Rubin, Intelligence Branch

Would you comment briefly on the question as to whether British intelligence failed to appreciate the ability of the Argentinian aircraft and weapons systems? Did they also fail to anticipate the ability of the British fleet to defend itself?

Elliot

This is one of the problems that is exercising the Franks Commission at this time. The intelligence community have been quite vocal in saying that they did advise the British government that things were likely to happen, but that the British government ignored that. I think that the Joint Intelligence Committee, which has its own priorities which are the direct threat to NATO and the ancillary threat to NATO from such places as the Middle East, puts everything else at the bottom of the barrel. They followed their normal pattern. Everything that came out of Russia was at the top. Whatever came out of Latin America took what time people were prepared to give it, which wasn't much. I am sure the air threat was as obvious to the British intelligence community as it was to commentators like myself outside that community. We were warning right from the very beginning of this scrap that the Argentine air was in range. Whether or not the navy thought they could handle all this is an operational matter, not intelligence. According to comments I got from navy people, they were quite sure that the Harriers could see off anything that came to them, and so they did. Whether or not they could handle it from the missile point of view is an open question.

General George Bell, CISS

Two questions of a strategic nature. First, how effective were national technical means in providing the force with intelligence about that part of the world's surface? And second, in the last part of your talk you touched on basing. Can you say what you see as lessons learned from this campaign relative to NATO out-of-area operations?

Elliot

There were problems. The British task force did not have the real-time intelligence for a lot of its field operations. I mentioned their lack of drones, which would have given them the advice that there wasn't anybody in Douglas. The commander just walked through the hamlet and kept on going. I think there was a glitch in the Port Darwin assessment. The original target there was two companies. That is why one battalion only was sent in.

There was a certain amount of operational intelligence I suspect - although I cannot prove this - from the American intercept station of Ascension. I think that information was going back to the United Kingdom and then coming down direct, on the scrambler, to the task force.

The question of out-of-area operations is becoming a political matter, rather than purely strategic. Consensus at the Ministry of Defence at the moment is that this was an aberration, something they don't intend to get into again, there being only about three places left where this would be a major consideration: Belize, perhaps Brunei, possibly Hong Kong. I think Gibraltar will be negotiated. I don't think there will be a fracas there. Their concept is still that our major threat is from the Soviet Union, and the question is, how do we handle the Soviet Union at its periphery? At its furthest extremity, that is the Persian Gulf. I don't think it will go any further than that, as far as Britain is concerned.

Bertin

It would seem that there was a complete failure to give real-time intelligence to troops once on the ground. Was this due to fundamental inadequacies in the equipment issued? Or was it because they hadn't been trained properly? How can we expect to cope in an extremely hostile electronic environment in Europe if we cannot even do it in Argentinian waters, out of range of enemy countermeasures?

Elliot

We have here a situation where a specialist force, designed to come ashore and secure a beach-head, which would rely for its intelligence on fleet-supported units, found itself in a longer term, longer range land battle. The only force which could be relied upon to do a solid job in bad country was the Marine and Para contingent. They are trained and equipped for this sort of thing in ways that the regular force are not. But they do lack the intelligence-gathering capability and the intelligence staff capability that the more regular units have. I think this was an aberration. There is training, there is experience, there is equipment available to the British army of the Rhine, to the Canadian forces, and to all the others in NATO. But it was not available in the Falklands.

The navy doesn't talk very much about their intelligence-gathering capability.

LCol L. Heaps

First of all, I'd like to know if the Canadians had any liaison with the British during this campaign, and if so, on what level? The second part is whether anything learned from the campaign could be or is being put to use by the sadly depleted Canadian forces, and if so, what those lessons might be?

Elliot

So far as I know, there were no Canadian liaison officers with the task force. There is a team in London which works very hard trying to keep abreast of what the British are doing in every branch of the service. As for what is happening at NDHQ, I'm afraid you will have to ask someone who has access to NDHQ. I do not.

Tim Descheneau, Systems Analyst

I would like to consider the use of armoured ships in war. During the last world war, one of the primary uses of battleships was for gunfire support, but also to draw fire from the carriers in a task force scenario. Given that modern cruise missiles tend to be optimized for use against unarmoured ships, would there not be a place for cruiser-sized vessels with guns and a little bit of armour?

Elliot

This is one that the Americans are trying to solve by recommissioning some of their four heavy battleships. They are having problems doing this. One of them is that when you fire the main armament, all the electronics go out.

The Exocet has been said, by one writer, to be equivalent to an 11 inch shell. Even the best of World War I and II armoured ships were vulnerable to an 11 inch shell, and that is still the case. One of the problems we all face in peacetime is the far higher standard of ship habitability than in wartime. Those of you who bounced around the North Atlantic in corvettes know that there were no amenities at all. Even hammocks were in short supply because you couldn't hang them anywhere. The modern ship has bunks and panelling and lockers; it is decorated with linoleum and gallons of paint, all designed to give the crew a home away from home. This is fine until the shooting starts. The ships are cleared out somewhat before going on an operational patrol, but if you tear something out you're going to have to put it back in and that's going to cost. The Treasury gets very unhappy when you add cost, so the tendency is to leave it there. My guess is that if we ever go to war again, there will be a few months during which ships will burn quickly and horribly, after which we'll go back to where we were in 1939-45, with a little bit of paint occasionally on the deckhead.

Descheneau

The Exocet is equivalent to an 11 inch shell in terms of explosive content but not in velocity. Several times when American cruisers and destroyers were fighting Japanese battleships at point blank range a shell, optimized to penetrate armour, passed through the destroyer, making an 11 inch hole, without exploding. If a cruise missile designed to pass through perhaps 40 mm of hull plate and explode in the interior of a ship were to strike two or three inches of armour plate it would make a large dent and a burn mark on the outside. If you employed a little bit of armour here and there, it would require an enemy air force to stock two types of missile.

Elliot

In the case of the Exocet that hit *Glamorgan*, the ship's spotters saw the missile coming at them and an order to change course was given and executed quickly. As she heeled away, her stern went down in the turn. The Exocet hit at about the rail, at an angle of 45 degrees and bounced along the deck. It burst just outside the door of the hangar. It wiped out the hangar, killed 13 and injured 14, but did no major damage to the ship control, communications or fire-fighting facilities. This would reinforce your point of view.

The difficulty, of course, is that if you put armour into a ship, you increase the weight. A lot of that weight has to be fairly high because of the weapons systems. This goes back to my earlier question: Do we bury the weapons systems? If we bury them, the crew will be hanging from the piping. This is the reason why they are putting aluminum into ships.

DAVID A. CHARTERS

PANEL AND FORUM: LESSONS TO BE LEARNED FROM THE FALKLANDS WAR

LCol MacDonald

Mr John Harbron will introduce our first speaker of the panel.

John Harbron

Professor Charters is at present the deputy director of the Centre for Conflict Studies at the University of New Brunswick. He is a native of Montreal and did his doctorate in war studies at King's College, University of London, on the subject of the end of the British mandate in Palestine. He has studied, written and lectured widely on terrorism, intelligence, special forces and covert operations.

Dr David A. Charters

There are two points worth making at the outset. The first is a general observation and it is simply that it may be premature at this time to draw 'hard and fast' *lessons* from the war in the Falklands. The principal reason for this should be self-evident: analysis takes time. It will take the British military, not to mention the broader western defence community, months if not years to fully digest the experience. British spokesmen have tended, quite sensibly, to emphasize the preliminary nature of their findings thus far. We would be well advised to do the same. Most of us are not privy to the classified analysis and the information available in the public domain - newspaper accounts and 'instant histories' - are wholly inadequate. Anecdotal and inconsistent, written without benefit of training in strategy, tactics and the principles of war, they must be used with care. In the meantime, one awaits the efforts of a good historian, but in view of the constraints of official secrecy, we may have to wait a goodly while.

A corollary to this first point concerns the use and misuse of 'lessons.' They easily can become 'politically hot' when they are used to justify (or to oppose) this decision or that strategic posture. Already, advocates and opponents of this or that weapons system have pointed confidently to the Falklands as *proof* of the correctness of their position. The 'big carriers versus small carriers versus no carriers' debate now raging in the United States is a case in point, but it also produces curious anomalies, such as the controversy over the use of aluminum in ship construction. In the rush to judgement and print, the uncritical eye can overlook the most important lessons, precisely because they are not those which seem most obvious.

The second opening point is more specific and concerns the land battle. By and large, it did not pit 'men against high tech.' It was, instead, a rather traditional infantry battle, fought in conditions reminiscent of the first rather than the second world war. The final clashes of arms were resolved by the bayonet, not the Exocet. That being the case, it may be that the most important lessons from the land battle are not those which focus on high tech, but which draw attention to the organization, training, leadership and motivation of men. I would urge those seeking meaningful 'lessons' not to lose sight of the 'human factor.' It may be the most decisive weapon of all.

Jeffrey Record, writing in the *Washington Quarterly*, has observed that:

> The most important ingredient of Britain's victory in the Falklands was the vastly superior quality of its soldiers, sailors and airmen and their cohesion under fire. The war demonstrated once again that the best weapons and most brilliant strategy and tactics are no substitute for training, leadership and ability to sustain unit cohesion under the stress and chaos of combat.[1]

He is not alone in attributing this to the regimental system which, he says promotes social and psychological bonds that sustain unit cohesion under the shock of combat.[2] This is hardly a new observation and the performance of the Second Battalion, the Parachute Regiment in the battle for Darwin/Goose Green seems to support Record's judgement.

But the fact of the matter is that the British had to rely on the qualities instilled by training and the regimental system to compensate for the limitations of *ad hoc* formations. Five Brigade, and to a much lesser extent 3 Commando Brigade, were composite formations as deployed to the Falklands.[3] The reshuffling of units prior to deployment was unsatisfactory, especially the deployment of two battalions (the Guards) whose most recent duties did not prepare them adequately for the rigours of the winter war. Moreover, 5 Brigade, by comparison, encountered fewer difficulties of this sort, even with the new units under command. And if there is a lesson to be taken from the British experience it is that *ad hoc* formations cannot be expected to perform as effectively as those whose structural integrity has been sustained for a long period. This is an endemic problem in peacetime because of the attention given to the career of the individual man, as opposed to concern for operational integrity of units and formations. Unfortunately, when war comes, the lack of cohesion and co-ordination can cost lives.

There is a further lesson to be drawn from this. But for the defence cuts in the mid-1970s, Britain need not have deployed any composite formations to the South Atlantic. The Labour Defence Review of 1975 disbanded the Parachute Brigade.[5] Although the three parachute battalions were retained, the brigade headquarters and the integral artillery, engineers and supporting services were lost. Consequently, although it soon became clear that the task of recapturing the islands required more than one brigade,[6] Britain did not have a second 'ready force,' in the form of a coherent, integrated all-arms formation, that could be dispatched quickly to support the Commando Brigade. The lesson here is that however commendable defence cuts may appear in peacetime, and often we have no choice about this, they constitute 'false economy' if they detract from readiness for war. Like *ad hocery*, false economy costs lives.

The most important lesson with respect to training and experience relates to professionalism: by this I mean the role of professionalism as a means of institutionalizing combat experience and its lessons. Long service, high standards of training and the regimental system permit both the men and the unit to develop an historical memory. This not only helps to motivate men and to preserve unit cohesion in the stress of combat, but allows commanders (from senior officers down to NCOs) to instinctively draw upon experience in order to solve immediate tactical problems.

This was most readily apparent in the British employment of raiding forces and amphibious operations. The Special Air Service, Royal Marine Raiding Squadron/Special Boat Sections were able to draw upon forty years of experience[7] in carrying out the missions assigned to them during the Falklands war: deep penetration patrolling, beach reconnaissance, target identification (for artillery) and raids on enemy installations (for example, the attack on the Argentine airfield at Pebble Island).[8] This is not to say that mistakes are not made. They are. But, on the whole, the ability of these units to integrate their historical experience, to translate it into tactics and doctrine that can be readily applied in battle, carries them through with a reasonable expectation of success.

Journalistic accounts suggest that the main landing was not simply an adaptation of an existing landing plan for the Persian Gulf, a possibility suggested by historian Alistair Horne. Nonetheless, the success of the landing at San Carlos was a tribute to the efforts of the Joint Warfare Establishment in keeping alive the skills, techniques and doctrine of amphibious operations over three decades, often in the face of government complacency or neglect.[9] The full story of the tragedy at Bluff Cove is not known, and I claim no inside insights, but it is worth noting that the troops involved were not experienced in amphibious operations.[10]

I note that a number of years ago, when the Crusader exercises were first started in Britain, there were a lot of jokes about the British army going to war on North Sea ferries. Well, they did and it worked. It is a tribute to the Joint Warfare people and the merchant marine that they were able to put together a reasonable amphibious capability from what was pretty much a scratch force.

Major Elliot made reference earlier to the problems of non-tactical loading. Given the shops that were available, I think the British managed reasonably well. That things would not go as well as planned was bound to happen. The phrase that comes to mind is: 'The plan of battle never survives the first clash of arms.'

At the same time it should be recognized that historical memory can be a problem. While the commitment to Northern Ireland provided invaluable training and experience for junior officers and NCOs, it left a legacy of 'minimum force' thinking and behaviour. Informed sources in the British military have suggested that in some instances during the Falklands campaign soldiers tended to react like policemen when what was required was real aggressiveness. If there is a lesson in all of this, it is that professionalism is essential to select and retain the important lessons of combat experience, so that they can be called upon when needed.

Some subsidiary lessons might include the following: first, the imperative of physical toughness. Winter in the Falklands exposed the troops to extreme weather conditions which reduced combat effectiveness. Temperatures dropped as far as -20° Centigrade; there were winds up to 60 knots, with constant snow and freezing rain. Moreover, the combination of weather, geography and sealift capacity limited substantially the number of vehicles available to the British forces. Mobility, therefore, was primarily by foot and, owing to the sinking of the *Atlantic Conveyor*, there was no tentage.[11] The corollary to the need for physical toughness is the importance of attention to the personal welfare of the troops – the soldiers were constantly wet and exposure and trench foot were universal problems. The Second Para was down to about 60 men per company by the very end, due more to trench foot and exposure than to combat casualties.

The second subsidiary lesson, but no less important, concerns infantry-artillery co-operation: infantry must be trained to the guns. They tended not to call upon the guns during night attacks until they were pinned down too close for artillery to be used safely. One battalion (I think it was the Scots Guards) was pinned down for more than four hours before calling for artillery support. Lack of training prior to the war meant that some platoon and section commanders could not remember the sequence of orders for requesting fire support and were not sure how to adjust fire. Likewise, gunner sergeants and bombardiers attached to infantry as Forward Observers (FOO) did not have a sufficient knowledge of company level infantry tactics and battle procedures. A further problem was that when the infantry did request artillery support, they wanted it to be continuous - and everywhere. The requested rate of fire was about 300-400 rounds per gun per day - about twice the normal rate.[12] Quite apart from demonstrating clearly the need for regular joint infantry-artillery training, it points as well to the importance of maintaining a vital link between the two - in the form of trained and designated artillery co-ordinators and FOOs in every infantry battalion.

On equipment and logistics, on the technical side, several items stand out. The first is the scale of the logistic effort required to support the land battle. One estimate suggests that, excluding ammunition (which was a major requirement in itself) and fuel, the British had to put ashore 25 tons of supplies per day, just to keep the troops fed and watered.[13] Ammunition consumption rates created a major logistic headache. 29 Commando Regiment RA fired more than 12 000 rounds from Goose Green to the final assault on Stanley, and it is reliably reported that some units had virtually exhausted their supply at the time the fighting ended.[14] Large ammo dumps had to be established, making severe demands on helicopter usage, given the shortage of tracked and wheeled vehicles, not to mention heavy lift helicopters.[15] The value of the helicopter as a resupply vehicle, and also for medical evacuation was proven once again, although apparently the tasking arrangements needed improvement to make best use of very scarce resources. The loss of the three Chinooks off the *Atlantic Conveyor* was a major problem.

The predominance of foot mobility, with a limited degree of air mobility meant that units were equipped to light scales and men tended to leave behind anything they could not carry on their backs. Artillery fire control computers were left in England; and other pieces of specialized kit - such as night observation devices, laser binoculars and radar units - were discarded in the rear areas as units moved forward, simply because they were too heavy. Similar complaints were apparently made about both the Carl Gustav and the Blowpipe.

Obviously, performance evaluations are not yet complete, but clearly deployment, handling and weather took their toll: ammunition got damp, missiles wouldn't fire, sights were broken. Major Elliot made reference to the moving of the fire-control units of the Rapier system up and down several decks through the ships before they were landed. All of this contributes to damaged systems and lower combat efficiency when they are put into action. The Rapier system performed well once operational, but took time to 'warm up,' thereby putting the troops at risk in the interval.[16] On the other hand relatively simple weapons, such as the 50 calibre heavy machine gun, performed with accustomed reliability. Fortunately, infantry brigades are combined arms formations and they do not depend for victory (or survival) on the performance of a single weapons system. What matters is the co-ordinated performance of all the integral and on-call weapons.

If there is a lesson in this, it may not be that high technology equipment is *ipso facto* good and necessary or unreliable and unnecessary, but rather that it has its limits. Those limits are determined as much by the shape of the battle and the conditions in which it is fought, as by the technical characteristics of the weapon itself. The fact that a certain piece of kit proved extraneous, or too heavy, or simply didn't perform as expected in the Falklands does not mean it should be rejected for use elsewhere. But it should force army formations to examine their equipment holdings and to determine, through rigorous testing and exercise, what is essential and what they can do without - if they have to. The war also pointed to the need for troops to adapt quickly to new equipment and to new scales of issue. Units which have trained for years with reduced scales encounter problems when vast quantities are made available upon deployment for war.

All of which seems to bring us back to the soldier, the 'human factor.' Lawrence Freedman, writing in *Foreign Affairs*, observed that 'the technical side of the conflict was once again shown to depend on competence and skill in using equipment, and a capacity for improvisation, as well as the basic reliability of the equipment.'[17] By fixing our attention solely on 'high-tech wizardry' we risk neglecting at our peril those traditional military virtues that can be decisive. For as Jeffrey Record concludes, 'War is still first and foremost a human encounter, and the intangibles of leadership, training, strategy, tactics and cohesion under fire are still as decisive as they were in the days of Alexander the Great.'[18]

Notes

1 Jeffrey Record, 'The Falklands War,' *Washington Quarterly*, vol 5, no 4 (autumn 1982), 46.

2 *Ibid*, 47; see also, Edward Luttwak, 'The Meaning of Victory,' *Washington Quarterly*, vol 5, no 4 (autumn 1982), 22.

3 At the outset of the crisis, 5 Brigade consisted of one Ghurka and two parachute battalions, without integral artillery, logistics or aviation. 29 Battery of 4 Field Regiment was affiliated for training, but was not on the order of battle. The two parachute battalions and 29 Battery were detached from 5 Brigade and deployed to the Falklands under command of 3 Commando Brigade. When 5 Brigade was called upon to deploy, it was 'given' two Guards battalions that had been on public duties in London, and a battery from 4 Field Regiment. Upon arrival in the Falklands the Brigade also took 2 Para under command again.

4 Sunday Times of London Insight Team, *The War in the Falklands: the Full Story* (Toronto: Fitzhenry & Whiteside, 1982), 249. See also Lawrence Freedman, 'The War of the Falkland Islands 1981,' *Foreign Affairs*, vol 61, no 1 (fall 1982), 207; and Bruce George and Michael Coughlin, 'British Defence Policy After the Falklands,' *Survival*, vol 24, no 5 (September-October 1982), 205-6.

5 The impact of the cuts is discussed in Michael Chichester and John Wilkinson, *The Uncertain Ally: British Defence Policy 1960-1990* (London: Gower, 1982), 154-56; see also David Greenwood, 'Defence and National Priorities since 1945,' in John Bayliss, ed., *British Defence Policy in a Changing World* (London: Croom Helm, 1977), 192, 204.

6 This undoubtedly is why the two parachute battalions were attached to the Commando Brigade in the first place. It is possible, however, that 5 Brigade was intended to serve as a garrison force, not for combat, but was the only other formation available once the task seemed too great for the reinforced Commando Brigade.

7 Useful recent accounts of the development of British raiding forces and doctrine may be found in Howard J. T. Steers, 'Raiding the Continent: the Origins of British Special Services Forces,' unpublished MMAS thesis, United States Army Command and General Staff College, Fort Leavenworth, 1980; and Glen St. J. Barclay, 'Butcher and Bolt: Admiral Roger Keyes and British Combined Operations, 1940-1941,' *Naval War College Review*, vol 35, no 2 (March-April 1982), 18-29. The post-war era is covered in Tony Geraghty, *Who Dares Wins: the Story of the Special Air Service, 1950-1980* (London: Arms and Armour Press, 1980); and James D. Ladd, *The Royal Marines 1919-1980* (London: Janes, 1980), 262-342.

8 The actual achievements of the British special forces remain shrouded in official secrecy and myth. Accounts of their activities may be found in Insight Team, *War in the Falklands*, 148-54, 185, 194-95, 198-201, 237; Brenda Ralph Lewis, 'The Falklands War - Recap,' *Army* (September 1982), 25-27, 29; and Edgar O'Ballance, 'The San Carlos Landing,' *Marine Corps Gazette* (October 1982), 27-39.

9 Alistaire Horne, 'A British Historian's Meditations,' *National Review* (23 July 1982), 887; Insight Team, *War in the Falklands*, 145-48, 185-93, 197-201; E. F. Gueritz, 'The Falklands: Joint Warfare Justified,' *RUSI Journal for Defence Studies*, vol 127, no 3 (September 1982), 47. Gueritz draws attention to the inherent contradictions in the 1981 Defence White Paper which, on the one hand, extolled the virtues of maintaining the specialized skills residing in the Royal Marines, and on the other, intended to scrap their amphibious assault ships.

10 For accounts of the Fitzroy landing see, Insight Team, *War in the Falklands*, 250-55; Freedman, 207; Max Hastings and Simon Jenkins, 'Tragedy at Fitzroy,' *The Observer Review* (31 October 1982).

11 The weather was consistently bad: freezing rain or snow on most days, temperatures as low as -20°C and winds up to 60 knots. The 'roads' on the Falklands (except in the immediate vicinity of Stanley) consisted of muddy tracks. Some 900 tents, sufficient for about 4 000 men were lost in the sinking, although the extent to which the extra weight might have impeded the advance tempers the loss.

12 Detailed, but undoubtedly selected, accounts of the employment of artillery in the Falklands may be found in *Gunner*, the regimental magazine of the Royal Artillery, no 141 (August, 1982), 17-20, and no 142 (September 1982), 11-17.

13 Michael Moodie, 'Six Months of Conflict,' *Washington Quarterly*, vol 5, no 4 (autumn 1982), 31.

14 '29th Commando Regiment RA in the Falklands,' *Gunner* (September 1982), 12-14; Freedman, 208.

15 Three of the four Chinook helicopters and seven smaller types were lost in the sinking of the *Atlantic Conveyor.* The remaining Chinook flew 150 hours, carrying 2 000 troops and 1.5 million pounds of cargo. See Derek Wood and Mark Hewish, 'The Falklands Conflict, Part 1: The Air War,' *International Defense Review* (August 1982), 980.

16 'The Falklands Conflict, Part 2: Missile Operations,' *International Defense Review* (September 1982), 1154; see also, *Manchester Guardian Weekly* (15 August 1982); and Edgar O'Ballance, 'Falkland Islands Campaign,' *National Defense* (September 1982).

17 Freedman, 207.

18 Record, 51.

LCol. MacDonald

Thank you, Dr. Charters.

I now call upon Commodore Hendy to introduce the second speaker of the panel.

NORMAN FRIEDMAN

Commodore R. I. Hendy

Ladies and gentlemen: Rounding out what might be called our triumvirate of experts from three countries, we now have a distinguished American commentator on naval and air matters to finish our morning programme.

Dr Norman Friedman was born in New York City. He took his Bachelor of Arts at Columbia University, and later, his Ph.D. from that same institution. He is a member of virtually every navally oriented organization in his country and elsewhere. He is deputy director of National Security Studies at the Hudson

Institute and has studied the United States-Soviet naval balance and the possible character of protracted nuclear warfare as well as possible Soviet reactions to United States naval innovations, both tactical and technical. He is particularly concerned with the impact of naval technology on United States naval tactics and strategy.

Dr Friedman has written extensively on ship design of all sizes of ships from carriers down. A resumé of the books and articles he has authored would take up all of his time this morning. He is authoritative, he is well informed, and it is a great pleasure to have him here.

Dr Norman Friedman

I think if you look at the Falklands, the overwhelming feeling you have is of the effect of political factors of various sorts, as well as of non-technological factors.

For example, the use of information on both sides was quite remarkable, and you can attribute some of the British successes to a very careful use of what we would call 'disinformation.' There was a great deal of publicity given, before the landing in San Carlos, to a policy of hit-and-run or land-and-leave. Such a policy would have been militarily completely insane, but it was believed by many, including the Argentines. That seems to have given the landing time to get ashore before there were any countermeasures.

Another major political point was the effect of inter-service rivalry on both sides. There is every indication that the three arms of the Argentine forces acted like three separate political parties in a coalition government, and that as a result co-ordination was essentially lacking. For instance, several people have pointed out that had the Argentine air force struck in conjunction with Argentine ground troops against the beach-head in the early hours, they would have beaten the beach-head off. The classic way to eliminate an amphibious assault is to stop it while it is happening. Instead, the resistance on the beach was not well organized and the Argentine aircraft had nothing to do with supporting the ground battle - at least, not for quite some time. Although Argentine air went in against the ships, the troops ashore were largely unbombed.

It appears to a naval analyst that on the British side a lot of the RAF operations were there largely to prove that air forces had some role to play, and not to help the battle in the least. It is very difficult for me to understand why it was worth sending a single Vulcan over to drop its 21 bombs on an airfield, supported by 10 (and some say 18) tankers! Later, when it turned out that exactly one bomb had hit the airfield, there was a background briefing in which the Operations Research people in the RAF explained that they had really expected only one bomb to hit! For reasons that I fail to understand, no one seems to have said, 'If you expected that, you should be out of business.'

Let us note the way British defence policy is made, the secrecy under which it is made, and the extent to which the British defence press (which is most of the English-speaking defence press) is uninformed about the nature of the weapons. For example, Sea Dart was advertised as the best missile you could buy; it was a ramjet, with what was said to be extraordinary range. Nobody mentioned that it had a hand-operated hoist or that it has a two-minute warm-up time. No other area anti-aircraft missile in the world that I know of that operates like that. That is a scandal. It is possible to get away with such things in the United Kingdom because defence policy is not the province of people outside the defence business; if you speak out they will squash you flat. Perhaps the problem of such a system of policy making is the main lesson of this war. That problem looks uncomfortably like what occurred in 1939 and 1940, when the main purpose of defence policy was to avoid embarrassment rather than to avoid losing.

One of the most interesting issues is that of estimate versus reality. The United States military believed that the war would be too difficult whereas the British admiral predicted a walkover. Both were equally foolish. The question is that when we look at warfare in general, or when we look at the possibility of a NATO war, are we being too conventional? Is our way of estimating what happens in combat terribly flawed because it is much too mechanistic? There has been a lot of talk about the training of the troops and the quality of the troops. But our ways of estimating what would happen in a war with the Soviets do not ever take those into account.

I must say, also, that the prevailing belief that you absolutely must have a naval base within a few thousand miles of where you are going completely belies the operational evidence of the Second World War. We built a navy to go out to the western Pacific without any bases. Oddly enough, most of the people who are doing the analysis in the Pentagon seem never to have heard of that. This is frightening.

On the question of political prediction, it was quite remarkable that within the governments there was no belief that a war would occur. Yet, as Major Elliot pointed out, it was quite evident that neither government could possibly back down. To us poor uninformed commentators, it was clear that there was going to be a fight, yet within Britain there was a belief that somehow they could reach a settlement. The Argentine estimates, I think, were mostly the result of a closed government and not of any belief that their weapons would work. That is important because, if you look at the chances for future wars, the Third World is a much more likely place than the Warsaw Pact, and Third World countries resemble Argentina much more closely than they do the Soviet Union. What does that tell you about the United States Rapid Deployment Force (RDF) in the Persian Gulf?

There was a lot of secrecy on both sides. It looks as though the Argentines did conceal their preparations well enough to prevent the British from pre-empting. They were so secret that the Argentine air force, which had the job of protecting those islands once they were seized, apparently had no say in the decision to seize them. So the fact that they were barely within air cover range was totally dismissed. The Argentine government was betting on a political resolution, not on tactics and not on technology. It is very important to note that military reality, which determined the outcome, did not determine the beginning.

We are told that this was an aberration. We are told that somehow operations in the North Atlantic will not resemble it. This stems from British governmental secrecy. There has been a belief in Britain that somehow, if you fight an anti-submarine (ASW) war in the North Atlantic, closer to the Backfires, Soviet naval aviation will not attack you. Somehow, it is not 'gentlemanly' to attack a British force with the worst area defence in the world (the Canadians are in fact even worse). It is believed that all of those Backfires exist simply to eliminate American aircraft carriers and that the United States will be kind enough to put our carriers up there to absorb them. The British government had plans for better area-defence ships and they cancelled them. No British carrier armed with Harriers has a chance of any kind against a serious Backfire raid. It does not have the numbers, it does not have the warning radars. The belief is that the RAF will somehow protect them.

Such decisions are political rather than technical. They are based on who had the best propaganda at Whitehall and who was best at manipulating the British government process. If you look at the British defence press, you will understand why it has no effect on these debates. I don't think that what I say here will have any effect there, but Canada is not Britain, and you should think very carefully about the role of independent organizations like this Institute in whatever debate occurs in Canada or, for that matter, in what happens in the United States. It is quite clear that the history of this war will be manipulated to show that whatever decisions were made beforehand were absolutely right. Whatever changes are made will be done very quietly.

Yet another factor is that this war is disproportionately well reported. It is being touted as one of the great wars of modern history, whereas a much larger war between Iran and Iraq is not being reported. That has largely to do with who is fighting. It does not have much to do with lessons.

As an aside I must say that as an American, I worry a great deal about the Rapid Deployment Force because I think the Falklands war was an example of exactly what the RDF faces. Again, if you look at politics, the RDF in its configuration is largely a function of inter-service negotiation, and not of lessons of warfare. The British were very fortunate in that they had a unified command and they were able to be reasonably realistic about who ran the battle. I'm not so sure that we would be as lucky.

The use of disinformation took place on a very wide scale. Most of the reporting during the war was well controlled by the British. The British showed that they understood the value of deception. Americans often have not. The British have a free press, but they were able at least to give out briefings that did them good. They also did very well on a more secret level. For example, they spread the word as an absolute secret within the United States that South Georgia was being taken for use as an airbase by Vulcans. Now, anybody who knows the geography of South Georgia will laugh that one off. But they knew that most of the people hearing this absolutely secret intelligence would not have the faintest idea what the geography of South Georgia is - myself included. It worked very well. I was very impressed that they were so good at it. They were good at it at all different levels. The business of protecting that landing was brilliant. It was exactly how you use disinformation, that is, it was used tactically. The reporters were furious afterwards, as reporters should be, but they had served their purpose. That gave the Task Force those eight hours that saved them from being wiped out on that landing.

Now, let us look at the weapons and find out how they did. Remember that much of what has been said has been directed, not to military lessons, but to the needs of the marketplace. First, there was Sea Harrier. It has many interesting characteristics, including the ability to manoeuvre under its engines in so-called VIFF. In point of fact, so far as I can tell, it used none of its capabilities. What the Sea Harriers did was orbit over the landing area. The Argentines came in on the deck because they were afraid of being shot down by missiles, and the Sea Harriers bounced them from above and behind, mostly with Sidewinders. (The Sidewinder has forward hemisphere capability, but it was not used so far as I could tell, even though it was relatively reliable.) They were doing look-down/shoot-down, except it wasn't with radar, it was with their eyeballs. It was very effective, but it was also very simple. All it proved was the fact that if you don't have a big carrier, you're stuck with Harriers.

The Harrier was a good air-intercept radar, but it has to be cued onto its target by someone else - and there was no one else. The British were in a very bad position there because they had scrapped their naval airborne early warning system when they scrapped their big carriers on the basis of a political decision. They looked at the possibility of putting airborne early warning aircraft on the *Bulwark*, which was originally a large carrier and which could have flown them off. But they had deliberately destroyed her arresting gear so that the ship could never be put back into service as a full-deck carrier. That was a consequence of their 1966 decision that there would be no more carriers, a political decision taken with tremendous enthusiasm. Well, if you run your defence that way, bad things happen to you afterwards! They have now announced they will do early warning with helicopters. But airborne early warning is a 24-hour-a-day job. If you really believe that a ship the size of the *Invincible* can fly those helicopters 24 hours a day for extended periods - good luck. And that's not even saying that the helicopter radar is awful - which it is.

Exocet is said to be the great ship-killer of all time. So far as I know, it never exploded when it hit ships except for the *Glamorgan*, and that was a fluke. When Exocet hit the *Sheffield*, what killed the ship was the fact that the missile passed through an oil fuel tank above the water line. It did not destroy the combat information centre in that ship; it hit her in the engines. If you put oil fuel tanks above the water line, you deserve what happens to you - as we have been told since 1905. The reason that the tank was there was because the ship had been limited in size politically. They were told to cut the cost. Ironically, you can make a bigger ship that doesn't cost a lot more - steel doesn't really raise the cost much. But because the British government was so obsessed with cost, and because it equated size with cost, it got the Type 42. It was a flawed design when it was originally planned. It was known to be flawed at the time. It is an all-steel ship. It wasn't the fittings that burned; it wasn't the toxic smoke. It was good old-fashioned oil fuel. We had plenty of oil fuel fires in World War II. They are very bad, and they aren't new.

When the Exocet hit *Atlantic Conveyor*, you learned the lesson that you should worry about protecting *all* of your fleet. The missile was deflected from its real target, the carrier. It searched for a new target, and there was this lovely large ship just waiting to be hit. That tells us that one has to think through the air defence of a group of ships as well as thinking ship by ship.

They are now advertising a vertical-launch Sea Wolf, nominally well suited to small ships and to converted merchant ships. But there are questions about whether it makes any sense at all at the range they're going to get. However, those questions will not be asked because the British have a controlled defence press and they are interested in selling. It is a very nice combination of a government industry and a controlled press.

If you look at Rapier, there was a question of reliability and a question of electronic interference between land-based Rapier and sea-based missiles. There are also apparently very serious problems about fire control measures. As far as I know, no one has explained exactly how they were able to separate the air-defence zone controlled by the Harriers from the air-defence zone defined by the weapons on the ships. It would have been embarrassing to have your planes shot down by your own weapons – but that happened quite often in World War II.

The French claimed that eight aircraft were shot down with Roland, but it is not likely that it shot down more than one – if that. Exocet also was credited with many more hits than it ever made.

In Britain, there is the major issue of the Conservative government preferring private industry to its own in-house design. One of their problems was that the Type 21 Frigate ('A' Class), of which they lost two, was a private design and apparently a rather bad one. But they can't quite admit that.

Was aluminum a hazard? It is not clear that aluminum really contributed to anything. Aluminum does melt. If you get a big fire, your ship may turn into a blob of molten aluminum – something that we discovered. Aluminum shatters, but you can do things about that. But if you buy a lot of radars you can put them up a lot higher in an aluminum ship. In the Type 21, aluminum was used because it was essentially an export ship for the Third World, in which the object was to show as many weapons as possible on the smallest (cheapest) hull – that's a commercial consideration. In point of fact, none was ever sold to the Third World, and the Royal Navy bought eight. But the design is typical of commercial types, which cut down on what one might call invisible factors that make ships survive in wartime. In the United States we endlessly have the problem that people look at our ships and say 'You don't have any weapons on them.' We say, 'We have enough weapons, but we also have the other things that matter.'

As far as habitability goes, it is very good for recruiting in peacetime. But go back to World War II and remember how the complements rose once you were fighting. Remember how many extra people and how much extra equipment had to be crammed in. That's when you are glad to have high decks and extra space.

There has been a lot of talk in British industry about how they could have solved all these problems if only the government hadn't been so 'backward.' I would argue that one of the problems the British had is that what has been bought has been determined never by operational considerations but by a combination of financial constraint and excuses. It is better to be honest about it and to leave spaces for equipment when you get the money, rather than to claim that all is well. Every British defence review reads the same way: 'We've cut it in half, but we've done even better because our commitments have changed.' But they haven't changed that much. And if the British are serious about helping us in the Gulf, they are going to have to go right back to this kind of amphibious operation, and they can't get out of it.

A criticism from outside observers about the Royal Navy and its handling of defence politics is that they don't know how to justify their mission, and they don't know how to survive in competition with the RAF. In general, air forces seem to be a lot better at political infighting than navies. Remember that inter-service warfare was a major factor in the Falklands war. The British had a single commander but the Argentines never did. I suspect, when the full story is told from the Argentine point of view, that inter-service rivalry will be the most fascinating part of it. The junta acted like a coalition government throughout the whole war, and never got its act together.

Resource management is a major issue in any war, both before and during. Why did the RAF need those Vulcan sorties and why did they need all those tankers? That was a terrible waste. It is not clear, by the way, that the RAF Harriers were worth the trouble either. But they had to be there so that they had a part in the 'big war.' Otherwise they would face severe political problems post-war. Similarly, I would argue that the Argentine navy sent in the *Belgrano* and made other attempts to sortie, not because they thought they would get such wonderful effects tactically, but because they knew that after the war, if the army were routed on the Falklands, they would have to show some lost blood as well. Otherwise, they would lose to the firing squad. If the army runs Argentina, and it feels it lost because the navy didn't show up when it was needed, there will be political revenge post-war. So I would argue that from the navy point of view, losing the *Belgrano* was a very good thing.

I would like to make some weapon comparisons, just to show what equipment is like. Sea Dart is roughly similar to the American Standard Missile. It has somewhat longer range, but the Standard Missile warms up instantly, whereas Sea Dart takes two minutes. The fact that you cannot keep Sea Dart warm in the magazine means that it is only good at very long ranges when there is lots of time to detect the target, pick it up and fire. At close range, forget it. Incidentally, that was never advertised, even in the classified literature, before the war.

Sea Wolf and Sea Cat have no exact parallel outside. The weapon that we use, Sea Sparrow, which is the same as the Canadian weapon, is semi-active and it can be argued that it is not very good at very low level, whereas Sea Wolf and even Sea Cat are command-guarded and ought to be better at low level. I think the reason that the United States will not adopt Sea Wolf is because it was not invented here rather than because of some defect in the system.

Sidewinder was common to both navies. The closest parallel to Exocet is Harpoon which is a much smaller weapon and probably more efficient. There is no real parallel to Sea Skua, although a helicopter-launched Harpoon would do the same job. The difference between Harpoon and Exocet is that Harpoon uses a liquid fuel, and if you believe that fire will kill most ships, then liquid fuel is what you want. Exocet has solid fuel. Solid fuel tends to shatter when it hits and doesn't do very much damage. My guess is that it was the sustained heat of the Exocet engine as it was passing through that ignited the oil fuel and started that fire in *Sheffield.* If you look at photographs of the undamaged side of the *Sheffield*, you will see no hole. That says to me there was no explosion, and I've heard that said by others. You could even argue that had there been an explosion, the fire would have vented and would have been easier to fight.

There was no British equivalent of an American carrier - as Americans like to point out. *Invincible* is closest to the American helicopter assault ships (LHA), and the new one, the LHD, will have a definite capacity to operate Harriers. There is no British equivalent of the American area-defence anti-aircraft ship. The Type 42, which is always touted as a destroyer of advanced technology, is actually closest to an American *Perry* Class frigate, which is the lowest end of our spectrum. The *Perry* has more fire-control channels, can fire faster and carries a lot more missiles than the *Sheffield,* (I hope that your new frigate when it is bought will be similar to *Perry*). *Sheffield* carried only 20 missiles. If you figure two missiles per target, that isn't very many engagements. The Type 42 has no point defence system, because it was built too small. The fire-pumps amidships, where the missile hit, had no auxiliaries aft because the ship was too small. The whole design was not realistic because cost came first, and nobody was able to say publicly, 'If you want to cut the cost that much - no ship!'

Type 22 is apparently a very effective ship and is closest to the American ASW frigates. Type 21, which is the inexpensive frigate of which two were lost (the 'A' Class) has no American equivalent, and I would suggest strongly there should not be any Canadian equivalent either because it is a disaster.

One of the points that comes up about the war is the incredibly high expenditure of ammunition, particularly ASW ammunition. We tend to cut down the number of weapons we buy as they become more expensive and more sophisticated. The British managed to drop about a hundred ASW torpedoes around Falkland Sound with no effect. The waters are relatively shallow, but the torpedoes were not designed for shallow water. There is now evidence that an Argentine sub did survive through that whole thing, but that its torpedoes failed rather than that the British succeeded in deterring it. I have even heard a report that the British kept it down for 36 hours and didn't realize it. That is very sobering for there are a great many submarines in the Third World, most of which are quite comparable to what the Argentines had, and there may even be some brave and intrepid submarine commanders out there. Apparently ASW, which the Royal Navy was supposed to be particularly adept at, just failed.

It also sounds as though satellite reconnaissance failed because of cloud cover - something we don't think about much. Reconnaissance was very spotty and the Argentines were able to fool it with mud craters on the air field. The British reported many times that they had knocked out that runway, believing what they saw in their pictures. When there are no real wars, the photo-interpreters always think they know everything and there is no counter-evidence. The lesson is, don't always believe what you see.

There were many unexpected points. First, the Argentines ran no top cover for those fighters going in. Had they done so, very possibly the few Harriers in the force would have been wiped out. A point not made is that when you only have 20-odd airplanes to protect your force, you tend not to waste them on ground attack or if you do, you tend not to be very aggressive about it, because God help you if you lose those 20 planes. They brought in another 20 during the war, but the total was still very small.

The Falklands war does not resemble 1995. It resembles 1945, only the numbers are much lower. It was like Okinawa, where the ships were tied to a fixed place, where the pilots had enormous élan and were quite willing to get killed. But in 1945, there were hundreds of defensive airplanes. Here there were at most 40. The defensive missiles, which in American doctrine would have made up the difference, were no good. You cannot operate with missiles which can't fire rapidly, with non-automated control. The British did not have any computers in the command centres (CICS) of ships like the *Sheffield*. There were very long delays between initial target detection and target engagement. The British radars are not well designed for tracking low-level targets, especially in land clutter. It has been reported that they had no moving target indication. That means you don't have a chance against aircraft coming over land. The entire Sea Dart system is notoriously unreliable and the programme to fix it was cancelled because it was too expensive – for after all, if you are never going to fight a war, why mention those little problems.

The *Sheffield* was a victim of secure communications and oversophistication. She was communicating by satellite. The reason you communicate by satellite is to make it difficult for someone to get a radio direction-finder (DF) fix on you. The Argentines obviously had no reconnaissance so they used radio intercept; thus the British used satellite communications. But when they used the satellite link, they could not use their countermeasures warning gear. It wasn't the radar that failed to pick up the Exocet; it was the countermeasures receiver which was blanked by the signal to the satellite. Every other ship in the force was able to pick it up and was jamming and firing chaff, but poor *Sheffield* did not get the message. I would guess that she was on silent picket, using that receiving gear to try to pick up someone coming in. But she was unlucky. I think the carrier nearby knew all about that problem (communications vs intercept) but *Sheffield* was a lot less valuable than that carrier.

By the way, there is evidence that the Argentine pilots were unaware of their

success when they fired, and the fact that the *Sheffield* stayed afloat as long as she did suggests that she could have been saved – if they had spent a little more money on fire-fighting gear and better hot-suits for the crew.

Here are some lessons from the Falklands. First, the vulnerability of surface ships. If you buy inexpensive surface ships that are defective – they sink. As for the capability of VSTOL, all we learned was that the Sea Harrier is reliable. It is very simple. You can fly it off a short deck. Its fancy manoeuvring ability was never tested because there was not much air-to-air combat. The planes they shot down could not manoeuvre. All that says is that if you fight Argentine planes 400 miles away from Argentina, it does very well. But in the Greenland-Iceland-United Kingdom (GIUK) gap, where the Russians are not gentlemanly, where they will send in Backfires and anti-ship missiles, I don't know what those Harriers can do. I have a terrible suspicion that what we have heard has more to do with selling Harriers than with real life.

As for the lethality of point-defence missile systems, we learned that you can shoot down missiles and other things with Sea Wolf. (Actually, I'm not sure it really did shoot down missiles.) Did they need a point-defence missile system? Did they need layered defence? If they had had a good area-defence missile in the first place, I would be more interested in the lesson. Since Sea Dart was so awful, I'm not sure I have learned anything about how other countries would behave. I'm not sure that I have learned about layered defence, and I'm not sure that I've learned that planes will always get through. What I have learned is that in a determined attack by 1955 technology against 1962 technology, the 1955 technology does pretty well. But it is now 1982.

The most interesting lesson is the value of a mobilization base. The British could not have fought without their merchant fleet. The United States has a small merchant fleet. If we go in somewhere, we may discover that merchant ships are still essential. This would be a great shock to us. I don't know what the size of the Canadian merchant fleet is or the size of the NATO merchant fleet, but I have my nasty suspicions. Nor can we believe that NATO will solve our problems, for there are a lot of places in the world outside of NATO concern. When was the last time, for example, you heard the Persian Gulf referred to as a NATO contingency? I never have.

As for limits of missile capability, Exocet is not very effective, nor is it really that difficult to stop if you know what you're doing. The iron bombs were quite effective. Because the Argentines came in very low, the iron bombs they were using often failed to fuse, and the number of duds was quite spectacular. An air force officer suggested that this was because they were fused for land attack and not ship attack. Had those bombs gone off, the British would assuredly have lost most of their fleet. In several cases, the bomb did not go off, but it touched something else off. For example, in one of the LSTs, the bomb landed in a pile of fire-lighters. The bomb did not explode, but the fire-lighters burned and the ship burned out.

What did we learn about Third World operation and maintenance? As far as I can tell, their airplanes worked but their fleet didn't. If you look at where those torpedoes apparently hit the *Belgrano*, it looks as if it sank because their watertight doors either were not closed or were not watertight and not because the torpedoes were particularly deadly. Ships of that type quite routinely survived worse damage than that in World War II.

What did we learn about the validity of the unopposed landing doctrine? In the United States we believe that you can always land somewhere the enemy isn't. But, if it is an island, you can't.

Inter-service problems seem to me to be dominant. If we do not address those problems in peacetime, inter-service rivalry will dominate our performance in wartime. It looks as if we are not serious enough now. Remember the comment about the impact of officers' careers. In a career system, rivalry between services will displace any feeling for a unified goal. That has something to do with the unreality of Western defence preparation in general, and is a very serious problem indeed.

What is the value of propaganda and deception? It is unbelievable. Yet we tend not to think about it.

On the validity of combat theories, our mechanistic idea of what combat is is bad. All is not what it seems and we were deceived almost constantly. Not only did the Argentines operate C-130s into and out of Port Stanley up to the end, but they also got a big container ship into that port just before it fell. That says that the blockade by the nuclear subs was not airtight – and probably didn't really exist. That says that when the weather is bad, all those Nimrods did not pick up anything. That ship was never attacked. It got in and got out and unloaded a large number of containers.

It appears that the Argentine fleet tried to sortie. It was not that they were cowardly, but their ships broke down – literally. They were World War II ships, not well maintained. The *Belgrano* was sunk by World War II torpedoes, not by Tigerfish. The original reports about Tigerfish were probably there to sell it. It looks as though Tigerfish is still a dud.

Notice the disparity in the number of RAF Harrier sorties versus Sea Harriers: 140 to 1500. It looks as if the land Harrier is not nearly as rugged. As for the role of the RAF bombers, I don't understand why those bombers were there. As far as I can tell, if they had all been melted the day the war started, nothing would have changed.

There are all sorts of stories about what the SAS accomplished. It is said that it destroyed all five Super Etendards and stopped the Exocet raids. False. A friend of mine was there and saw them after the war ended. They predicted the raids to give early warning. But I question whether it paid off that much. I don't really know. They have a great tradition of absolute secrecy. It probably is a very intelligent thing to do.

Politics remains a very serious problem. The British government decision to cancel the carriers and to cancel the air-defence ship which would have followed the Type 42 was based on the claim that there was no air threat in the eastern North Atlantic and that anyway the RAF would take care of it in a general European war. I don't understand how they got away with these claims. The Soviet navy has a very large, active and effective air force. We worry about it in the United States. If we have to worry about what those airplanes can do to a carrier battle group which has dozens of times the anti-air capability of a Type 42, I don't understand how the British can believe that Soviet air is not going to show up. If every lesson of history shows that land-based aircraft will not protect you at sea because they are not available when you need them, how can they believe that Nimrods and Tornadoes will protect them in the North Atlantic? In a general European war, who can imagine that those airplanes will not be sucked into that cauldron in central Europe? The question is, why didn't anyone ask about it? Why was there no scepticism in the official press? How did they get away with it?

The naval Battle of the Dogger Bank (1915) is interesting as a historical parallel; it was the first case of major ship combat in World War I. The British performed very poorly tactically. Their signals were a disaster; they failed to destroy most of the German force. But one of the German ships, an obsolete armoured cruiser called the *Blücher*, was sunk. The next morning the British newspapers carried a large photograph of this ship turning turtle. But they did not ask the question, 'Why did the British battle cruisers fail to sink anything else?' There was no enquiry. In particular, the officer responsible for the garbled signals suffered no consequences. Nor were serious material deficiencies in the British ships remedied. As a result, the next year there was a larger battle, Jutland, in which several British battle cruisers were lost.

I worry that the success of this war will engender so much self-congratulation that the problems it raised will not be made clear. The war is being called one of the greatest of all time. The performance of the weapons is being advertised very heavily. Losses are being called 'remarkable, but necessary.' There was no reason to lose anything. There was no reason for *Coventry* to have been lost through saturation of her air defence. There was no reason for *Sheffield* not to have been saved. But I will bet that *Sheffield*'s captain comes out of this with a decoration and that *Coventry* is put down as an unavoidable loss. Too bad for us.

LCol Macdonald

Thank you, Dr Friedman.

At this point we invite comments and questions.

PANEL DISCUSSION

MGen Robert Larose

My question concerns the vulnerability of helicopters. In the Canadian forces helicopters are deemed to be too vulnerable to go forward of land troops. Their task in reconnaissance is rather limited if all they can do is go to the level of the front patrol cars or patrols on foot. Are attack helicopters also limited to the front edge of the fighting troops?

Elliot

The British experience, although the evidence is not complete, seems to suggest that the real value of the helicopter (with all the limitations they faced in the Falklands) was in resupply and medical evacuation. They did put helicopters forward to do reconnaissance on some occasions.

The real question you must ask is, 'If I put my helicopters forward, knowing that they are vulnerable to some extent, how many helicopters, whether attack or observation, and their crews can I afford to have shot down,' because sooner or later your resources are going to run out. That's the best answer I can give to you. Helicopters in the Falklands in the land battle were used primarily in re-supply: moving artillery and carrying a lot of ammunition forward to supply the very high consumption rates of the artillery.

Unidentified

Bob, I'd like to pick up this point about the losses.

There was one scout helicopter lost at Darwin to ground fire. It was forward doing a combination of reconnaissance and artillery fire spotting. The two

Gazelles that were lost at Carlos were again picked off by ground fire. They were down low. I'm not sure whether they were picked off by gunfire or by Blowpipe. They were too far forward and got involved in the support weapons of the small contingent of Argentinians, about 40 in all, which was left in that battlefield area.

To pick up another point: there was no opposition on the ground by the Argentinians to the landing itself. The Argentinian commander got himself caught in the trap of keeping his major force back too far to be able to counter a landing, and he did not have the mobility or the helicopter support needed to move troops forward to pinch off the invasion by the British. The counterattack was, in fact, an air counterattack and very clearly that attack was wrong; it went against the wrong targets.

Charters

The remaining Chinook (that was not on the *Atlantic Conveyor* when it was sunk) performed extremely well at a very high rate of utilization, carrying something in the neighbourhood of a million and a half pounds of cargo and 2000 troops in about 150 hours of flying, even after losing its navigational equipment, a cargo door wheel and its windscreen wipers. That says something for the performance of the Chinook helicopter.

LCol J. Cameron

I want to make a short comment on a technical matter that arose regarding the question of missiles against armour. Missiles and the various designs of their charges will defeat any steel armour that is produced. The argument for putting more armour on ships to defeat missiles is of no value. The various types of charges, whether it is heat, hesh or other combinations of chemicals will defeat any steel armour that's designed. What is required is different types of material in combination with electronics. The armour itself is simply not protection.

Nicholas Stethem, Strategic Analysis Group

I just returned from ten days in the United Kingdom and I got some details which may be useful.

First, the captain of the *Sheffield* did not get an award. It was the gossip of the mess that he was in up to his neck because there is something more to what happened there than has been reported. I don't know what it is. Second, the Harriers did use the motors in order to do the VIFFing. Some of the Argentinian aircraft were refuelled, so the Harriers did have to dogfight. VIFFing is a simple, not a complex, thing.

In a dogfight, when the enemy turns in order to dive down on you, you go into a tight turn, and fire your motors out – which pulls you around really tight, so that as he's coming down on you, you're coming up on him. The Harriers had been practising that tight turn, and apparently it works effectively if the other pilot doesn't know what you're going to do.

The difference between the naval and air force Harriers, according to the pilots, was the Blue Fox radar. The air force Harriers didn't have it, and were of limited use.

According to the Blowpipe crews, and I talked to a couple of corporals who were Blowpipe detachment commanders, the hardest things to hit were the Puccaras. The second hardest were the helicopters. They had to be pretty much direct on. Right now the British are exercising on Salisbury Plain with a locating system for Blowpipe detachments so that you can catch jets coming in at different angles.

Also, I watched Tornado beat Rapier consistently by using the terrain-following radar. When they used the radar to bounce over ridges at low height, Rapier never had time to catch them. Whether they would hit the target is another question! But they were exercising this.

According to some naval officers, the burning of the *Sheffield was* made worse by artificial materials, including the electrical cable which turns into some sort of noxious gas when it burns.

There were many other things. I talked to the Ops Officer who had done the assessment on why the armour was not sent forward initially. It was a mistake on their part. The problem with ammunition was with the time taken to get it forward because of the limited number of helicopters. The Argentine forces could thus regroup and the British would lose momentum. The question of food was the same. The reason they did not use helicopters to transport men was because of a lack of helicopters.

The people carrying Blowpipe (and some carried Milan, too) made the point that just about everything they had was designed for mechanized warfare and to move no more than 100 yards from an APC. It just didn't work for them.

You say SAS and SBS operations are very secret. But I talked to Roy Walker about it and, apparently, the key there was reconnaissance – giving your own troops that edge when you were going in or having the SAS there on the ground, doing something nasty in order to help you. In other words, it wasn't the dramatic on-the-coast stuff – although something did happen over there but I wasn't able to find out what – but the use of men on the ground far, far forward on the island itself.

Friedman

The issue of plastics aboard the *Sheffield* is an illustration of the reality that the perception of the operators on the scene is extremely different from that of the planners and constructors. I come from the planners and constructors end. When you say that the operators should make design decisions you have to realize that often their perception of what they are seeing may be defective. There was a fire in bedding on the frigate *Plymouth*, which was apparently the closest to a pure case of plastics burning and they found it wasn't very serious. Yet it was notorious within the service that plastics had somehow contributed heavily to the loss of the *Sheffield*. A good solid oil-fuel fire will do terrible things, as we learned in the south Pacific.

John Harbron

There is very little PVC in the wiring. The biggest problem was that the cables dropped down in the companionways and they couldn't get through.

Friedman

It sounds as though the damage control was very bad and hadn't been practised.

Charters

Another comment that I heard was that the galley was functioning at the time and there was a lot of food grease which contributed to the spreading of the fire through the galley area. How much of that is true, I don't know. But it could have been a contributing element.

Major John Hasek

One thing I haven't heard about is the Argentine troops on the ground. Surely, if one thing lost the war for the Argentines it was that in the end their troops lost cohesion. Had they not done that, there is no way they could have been beaten because apparently they had armour on the ground and lots of artillery. I think it was British propaganda that managed to win that war. Without that, nothing anybody could have done would have turned the scale. If you look at logistics of supply, and the long lines of communications that the British had, surely the thing for the Argentines to do was to sit in Port Stanley and keep extending the British, sucking them further and further along. How could the British have taken Port Stanley if the Argentine troops hadn't broken?

Friedman

The lack of cohesion of Argentine troops was not a function of British propaganda, but had to do with the way the Argentine army was organized. The command of the islands was under an intelligence officer. Have you ever heard of an intelligence officer being in command of a battle? He got that command because he was loyal to the regime. The officers were reported to be enjoying themselves in Port Stanley while their troops were freezing up front. That's typical of Third World armies. That's why they break. It's not that you tell them that the war is hopeless. It's because most of those armies are hopeless. Giving them all the goodies in the world, which they had in great supply, has nothing to do with their fighting quality. If you talk to anyone who fought the Egyptians in 1956 and 1967, they will say the same thing. They would have been great troops if they had been well led. So far as logistics go, talk to whoever commanded in Okinawa or Saipan on the Japanese side. They had lots of troops, lots of stuff. But they didn't win, even though we had a very long line of supply. You can move things 10 000 miles by sea a lot more easily in many cases than you can move it a couple of hundred miles by bad roads on land. That is why invasions by sea have often done extremely well. We keep forgetting that. You don't have to have a base there – just bottoms.

Charters

There were some cases, particularly in some of the final battles, where some Argentine units and sub-units actually fought very well. Nonetheless, I think I would agree that the whole problem of leadership of the Argentine army was a vital factor in their defeat. They were, in fact, better equipped with night-fighting equipment than the British. Some of their cold-weather gear was better. But they didn't really know how to use it. Some of their tactical procedures were very lax. They didn't have outlying pickets around their positions, so the British were able in many cases to get fairly close before starting the attack. Their counter-battery artillery was very poor and was not co-ordinated.

Elliot

I think we should remember that the original operation from the Argentine point of view, was very largely bluff. They hoped that they would create a situation to which the British would not respond. When the British did respond, they themselves had to try to draw another card. The card they drew was a very large increment of troops from the Southern Command. Most of these were youngsters who had just come in on the conscript draft. They were not really past basic training in most cases. When they got ashore, they were faced with a situation that would have strained seasoned, experienced combat troops. They were not taught to work as units. They were not taught to work with artillery. They were not taught the proper logistic and command support functions which are normal in any reasonable organization. This is the chief reason for the defeat they suffered. They were not properly trained for the job. The troops that *did* fight were the Bujwa Tactical, the Marines, and the few regulars who were in that original task force of 3000 men that came ashore and the immediate follow-up to it. That must be remembered when we look at the casualty rates, and the way they broke and ran after the outside defences were penetrated.

There were errors, as I mentioned, in the actual tactical handling. You may argue, as Dr Friedman did, that they did not counterattack. More serious was the fact that they took the troops out of Mount Kent and Mount Challenger, the outposts of their defensive horseshoe, and moved them to Darwin, at the very best a side issue. They should have, in fact, vacated Darwin, and brought everybody back into the horseshoe. They should have pulled people out of West Falkland and put them into the horseshoe, if they intended to hold Port Stanley as a long-term base. The British would not have been able to take Port Stanley against a reasonable, proper defence.

Dr L. J. Calvert

I'm a medical doctor and I was very interested in your report on the incidence of trench foot in the troops who landed and fought such a short battle. It seems a very short time for them to get trench foot in that number. The only cases that I have seen comparable to that was the incidence of gonorrhea in the First Canadian Corps in Italy in 1944. Serious steps were taken to prevent that going further, especially against bringing it back home. Shouldn't serious steps have been taken here? In the Second World War, trench foot was almost considered a self-inflicted wound. It seems incredible that with the supplies we have now, and with the foot gear and clothing, that so many cases of trench foot should occur. It sounds as if those boys went ashore in sneakers so they could sneak up on the Argentines. If the Argentines had been smart, if they hadn't fired a shot and just sat back and waited till the whole of the expeditionary force had gone down with trench foot, they might still be in possession today.

Charters

That's a real wild card question!

I quite agree with you. Somewhere along the line there must have been a failure of equipment, particularly the boots. I don't have evidence to support the assertion that the combat boot was a bad one. We saw a medical report from 2 Para that listed the casualties from trench foot and from exposure. It may be that the British land forces were not as well equipped for the weather extremes that they encountered. Perhaps they wanted to push forward too quickly and simply sloughed off what they considered to be minor problems, so that the soldiers were told not to complain if their feet were wet, and that sort of thing. I don't have an answer.

BGen C.L. Kirby

My comment, as an infantryman, is to Dr Charters. I hope that the lessons of the Falklands will go beyond the Falklands, which is probably the last example of imperial policing we'll ever see.

The trouble with small professional forces when they are involved in large wars is that they get overrun by large, mobilized semi-professional forces. The problem is not how well small professional forces can apply the training that they have learned over and over again for forty years, but how well they can transfer those lessons to large mobilized forces to provide them with backbone and knowledge in a big hurry, and make them able to stand up against a large professional force such as we face in NATO. I'm terrified that the lesson of small professional forces will be clutched by those countries that maintain small professional forces and have no mobilization plans and used as a crutch for that very shortsighted and unprofessional view of how future wars will look. One further point: small professional forces are not necessarily efficient: many of them are just small.

Just a couple of points on lessons learned and lessons unheeded. The *Belnap* was an American cruiser with an aluminum superstructure. She ran into an aircraft carrier in the Mediterranean several years ago. Her entire aluminum superstructure was burned and destroyed. The only thing that survived was the steel hull. I wonder how that lesson has been applied in the four or five years since the *Belnap* incident.

As an example of lessons learned but unheeded in Canadian terms, I will give you two on winter equipment which have a bearing on Dr Calvert's remarks. We have a piece of footgear for the army, produced in Quebec, which is probably the worst designed piece of footgear ever issued to any army. In the first place it takes about five minutes to get laced up, and that's not very useful when you've been asleep in a trench. It flays the feet of infantry, even well trained infantry. Reports on the anti-soldier effect of this boot have been turned in over the course of several years. Also, our winter clothing (which the British adopted and I hope have given up) is designed for the high dry Arctic. It acts like blotting paper if you go to Norway where we have a brigade group committed, and would have acted in exactly the same way in the Falklands. When I commented on this as the commander of CAST combat group, over and over again, it was met with the kind of response that Dr Friedman so capably demonstrated. That is, because it was a saleable item, any comments were unwelcome.

Mr V. Upans

Dr Friedman, you have directed most of your comments this morning to the lessons NATO should have learned on fleet air defence. I wonder if we could look at the opposite side of the coin. It would seem that notwithstanding some hundreds of sorties, only half a dozen British vessels were actually sunk. What lessons can we expect the Russians to have learned from these experiences in the attack mission, either from the point of view of doctrine or technology?

Dr Friedman

The Russians have written it up and have said that saturation pays. Soviet doctrine is to try to get as many weapons on target as they can and they use large anti-ship missiles to do it. If you look at the number of ships hit rather than the number sunk, many British warships were hit, but the bombs dudded. Had those bombs exploded, it's clear that the ships would have been lost. It's difficult for the Russians to learn lessons on the vulnerability of ships because most of the ones they want to sink are much larger than the British ships that were damaged.

In a type of air defence, in which ships are not allowed to manoeuvre, you can expect to lose about a third of your aircraft each time over. If you are going in with weapons, each of which is large enough to sink, then that is not such a terrible loss rate. While the air force becomes rapidly demoralized you can sustain it for a little while. Saturation has a big payoff. But you can't really tell from the Falklands war, because the weapons they were up against were second rate, and not the ones they would go up against for real.

GEORGE LINDSEY

THE HIGH TECH BATTLEFIELD: EQUIPMENT REQUIREMENTS INTO THE 1990S

LCol MacDonald

Ladies and gentlemen: During this morning's session we had an opportunity to carry out an environmental analysis from the standpoint of a recent operation. This has now passed into history. It represents the environment of six months ago. The focus of this particular seminar is on the future environment to which we must adjust, with all of its technological capacities and its military implications. This afternoon we shall start with a detailed examination of that future technology. General Leonard will introduce our first speaker.

General William Leonard

Ladies and gentlemen: We have had a stimulating morning. I have attended eleven of these seminars out of the twelve, and I don't think I've heard a more interesting series of lectures than I heard this morning. Of course, anyone who tells me that the regimental system is great has me in his pocket right from the beginning.

We have had a peep through the window at the war of the 1980s, and some indications of the 1990s, in the three environments and all this crowded into 45 days. It is the kind of thing, set in perspective, that is an operational analysis organization's dream.

This afternoon, George Lindsey is going to look at this peep through the window from the point of view of the chief of the Operational Research Analysis establishment for the Department of Defence. George is going to tell us something about how he will analyse the impact of the technological environment, the high tech battlefield, on defence equipment requirements for the 1990s.

Dr George Lindsey

CONFLICT IN THE 1990s

Some Working Assumptions

In order to discuss 'The Battlefield of the 1990s' it is necessary to make some assumptions about what will happen during the 1980s, and also to decide whether our battlefield is a nuclear one or not.

I have assumed that mutual nuclear deterrence has prevented the outbreak of World War III to the end of 1982, and promises to remain mutual and reasonably stable through the rest of the 1980s. Let us also assume that the deterrence which is produced largely by the presence of nuclear weapons will continue to apply into the 1990s, but that during the 1990s international conflict will reach the stage at which large forces of well-trained ground troops, supported by tactical air forces, and equipped with the latest weapons of the day, will engage in operations of war. Assume furthermore that, although the threat of nuclear weapons is present, the war does not go to the nuclear level. Otherwise, we would have to talk about something else.

The Battlefield

The word 'battlefield' suggests a place where there are well organized operations between clearly identifiable forces in close contact, in an area allowing for manoeuvre. There are, of course, many other types of situation in which military conflict can occur, such as guerrilla warfare, counter insurgency operations, or forcible imposition of economic blockade. But the high technology battlefield of the 1990s will not include that sort of operation. The maritime element would also have to be treated in a proper discussion of the 1990s conflict, but in order to finish within my time, I will talk only about the land/air battle.

Another thing one must do is to decide just which weapons systems are going to be around in the 1990s. It is important to remember the long time-cycle through which a weapon has to pass before it is deployed.

Probable Developments in Battlefield Weapons by the 1990s

To have an important effect on the battlefield, new weapons must have completed their research, development, test, and evaluation phases, gone into mass production, been accepted in quantity by the armed forces, and been deployed to the operational formations. In addition, doctrine and tactics to exploit their new characteristics must have been evolved, and it may have been necessary to modify force structures. Finally, the troops must have been trained to use the new weapons. This sequence of events takes many years, especially in peacetime. Consequently, weapons and equipment still in the earliest stages of the sequence at the end of 1982 may not have much influence on a battle occurring in 1990, although they could before 1999.

Unlikely Developments

One encounters forecasts of laser or death-ray weapons likely to revolutionize military combat. There is no doubt that great advances are being made in laser technology. Already there are very useful applications of lasers in the roles of range-finding and of target designation. These use very little power, and are instruments rather than weapons. We can expect further developments before long for such purposes as surveillance and communication. However, to become a weapon of destruction it will be necessary to deliver large amounts of energy to the target. Not very much energy would need to be delivered to injure an unprotected man, considerably more to damage an aircraft or thin-skinned ground vehicle, and a lot more to damage a tank. The equipment needed to generate the necessary power and convert it into a laser beam may prove to be bulky and immobile, and it is doubtful that many practical laser weapons will be deployed on the battlefield in the 1990s. You might see them on a battleship or in a static location. But I am going to assume that practical laser weapons, as opposed to instruments, are not something that we will have to worry about by ten years from now.

Great progress has been achieved in the use of orbiting satellites to observe activities from space. Further improvements are to be expected before the end of the century, especially in the detection and tracking of aircraft in flight. Such capabilities will be increasingly important for intelligence gathering and strategic warning. It is not, however, certain that the data handling capacities or the ability to detect with high resolution under conditions of poor visibility will enable space detection systems to play an important part in the tactical conduct of land battles in the 1990s. As Dr Friedman pointed out, cloud cover can vitiate the effects of a photographic reconnaissance by satellite, and I think information overload might well make it difficult for hard-pressed commanders to use the information.

Cruise Missiles

Although cruise missiles date back to World War II, and have been developed by the Red navy for very effective naval use through the 1960s and 1970s, recent spectacular advances in the technologies of guidance and of propulsion offer many new applications. An advantage of a cruise missile is that once it has been developed, it is comparatively easy to arrange for its launching from a ground vehicle, an aircraft, a surface ship, or a submarine. There is great flexibility of employment. The mode of guidance best adapted for accurate navigation over land is by terrain comparison radar, comparing what it sees on the ground with a map. This requires the accumulation of precise digitized contour maps of the territory over which the missile will fly, right to the area of the final target. That would mean a great mapping exercise over a large part of the world before the war starts. The best way to construct such maps is by special radar satellites, a realm of technology at present only possessed by the United States and USSR. So it's not a simple thing that anybody can use on short notice.

The most newsworthy applications today are the ground-launched cruise missiles intended to carry nuclear weapons to strategic targets in Europe and to strike surface ships at very long ranges. However, their technical capabilities and comparatively low costs are such that they may acquire roles on the battlefield of the 1990s, armed with conventional warheads and quite possibly replacing aircraft for some functions, and extending the range of air-delivered weapons for some other functions.

Cruise missiles are likely to fly at very low altitudes at subsonic speeds, or at high altitudes at hypersonic speeds. They tend to be smaller than aircraft, and so present more difficult targets to radar. But defence against the cruise missile is not impossible, especially on a modern battlefield already well equipped with anti-aircraft weapons. We should not consider it as something that can never be stopped.

Ground-based Air Defences

Although anti-aircraft guns of yesterday could fire shells that moved much faster than their airborne targets, the difficulty of predicting the motion of the aircraft was such that AA fire was notoriously wasteful, especially when the target flew at high altitude. The answer to this deficiency was the surface-to-air guided missile, able to alter its flight path in order to conform to the motion of the target. A missile can easily be made more manoeuvrable than an aircraft (one reason being that it doesn't have a man inside it), and can be made to home on radar or heat signals from the aircraft.

By the 1960s and 1970s surface-to-air missiles had got the measure of aircraft flying at high or medium altitudes. An example was the famous destruction of the American high altitude U2 reconnaissance aircraft by a Soviet SA-2 missile over Sverdlosk in 1960.

Because the larger surface-to-air missile systems require some time for the radar fire-control to acquire the target and lock on, and for the missile to accelerate, they are at a disadvantage against targets moving fast at low altitudes. It is against this type of target that multiple AA guns have a better chance, especially if controlled by radar, automatically slewed, and if they have a high rate of fire. Also, against low-altitude fixed-wing aircraft and helicopters, which generally fly rather slowly and at low altitudes, AA guns and the smaller shoulder-fired heat-seeking SAMs can be deadly.

As a result of the increasing effectiveness of air defence, aircraft will not be able to operate above the battlefield of the 1990s until they have succeeded in suppressing these ground-based defences. The two main methods of accomplishing this are by countermeasures, electronic against the fire-control systems, aided by timed manoeuvre, and by physical attack, whether from the air or the ground, of the missile sites and control devices.

Aircraft

For seventy years after the first flight, the designers of military aircraft engaged in a race for greater aerodynamic performance (that is, speed, altitude, and range). Today this race has virtually ended, with the military value of an aircraft far more dependent on its armament and its avionics than on the airframe itself.

Several good basic air frames, such as those of the A3, F4, and MIG-23, have been adapted for various roles, and with the fitting of external pods it is possible to equip one aircraft to perform different tasks in a matter of minutes.

There is one counter-trend, though, against this multi-role use of aircraft. Because of the increasing effectiveness of ground-based anti-aircraft defences, we have seen the emergence in the United States of the A10, a slow, armoured aircraft carrying a heavy load of air-to-surface weapons. It is a specialist in the close support of ground troops. Beyond this, remotely piloted vehicles (RPVs) are appearing and may replace unarmed aircraft for some of the most dangerous roles. I am sure we will see more replacement of manned aircraft by these vehicles by the 1990s.

Large air bases and aircraft on the ground are now so vulnerable to attack by air or by missile that it is doubtful that they will be able to survive long enough to support operations for any length of time. Accordingly, great advantages are obtained by the use of vertical and short takeoff and landing aircraft. We heard a lot about the success of the Harrier this morning. In the Falklands it was being used in its VSTOL aspect until the British were established ashore because it could take off from the small deck of a ship. But the aspect I am speaking of is the dispersal of aircraft. You could hide them in the trees and fly them from very small pads or even highways.

The operation of fighter aircraft which wish to intercept enemy bombers, and of friendly attack aircraft which wish to avoid enemy interceptors will be greatly aided by Airborne Warning and Control (AWACS), which can oversee the enemy side of the forward line of troops, track targets at all altitudes, identify hostile aircraft, and direct engagements by friendly aircraft and ground defences. We heard how the lack of early warning was a severe disadvantage for the British fleet. It would be an equal disadvantage for forces operating on land, trying to keep enemy air off their backs.

To conclude our discussion of aircraft, there will be increasing dependence on helicopters for a multitude of applications, including attack as well as transport, but it must be recognized that they are extremely vulnerable to ground fire. That is one of the great limitations of helicopters.

Aircraft Armament

More important than the airplane is its armament and avionics. Very important improvements are being made to airborne radar, including Sideways Looking Radar (SLAR) which can map features on the ground with high resolution, and Airborne Moving Target Indication (AMTI) which can detect and track aircraft flying at very low altitude. That is where the look-down/shoot-down capability will come from.

We have heard about 'smart' weapons. They were much in the news in the later stages of the Vietnam war. Airborne weapons are getting 'smarter' all the time, so that it is no longer necessary to assign a large number of aircraft to the sure destruction of one key target in order to overcome limitations in the accuracy of bombing, although it might be necessary in order to ensure that one or two aircraft manage to penetrate strong AA defences.

The principle on which some of the most successful 'smart weapons' operate is homing on a 'spot' marked by a laser designator beam. This beam can be steered by a forward observer on the ground, by one of the aircrew in the launching aircraft, a helicopter or another aircraft, or possibly by an unmanned vehicle under remote control.

The vulnerability of airfields is arising partly because munitions are being specially designed for the attack of airfields, to crater runways and to create hazards to personnel attempting to repair the damage. These depend on penetrating weapons that will go through the crust of the runway and make a big explosion underground that will create a large crater. Another unpleasant aid to attackers of airfields is the dispersal all over the airfield of small booby traps which would make life very difficult for the people who had to go out and fill in the crater.

Recently developed missiles now make air-to-air engagements possible at very long range, provided that the aircraft is equipped with a powerful radar. The F-14 Tomcat, a United States navy fighter, is probably the leading example. However, identification is notoriously uncertain until visual contact has been achieved, and at shorter ranges the advantage may be with manoeuvrability and cannons rather than with sophisticated missiles. This has been verified in both exercises and real combat conditions.

Air-to-air missiles can home on radar or on heat signals, and both are subject to countermeasures. Homing can be difficult against targets flying at very low altitudes, although that problem can be overcome with modern techniques.

Anti-tank Guided Missiles

During the 1970s, the first generation of anti-tank guided missiles using 'manual command to line-of-sight' guidance reached a high state of perfection. These required the operator to steer the missile all the way to the target, usually by instructions transmitted through a pair of trailing wires. That required the undivided attention of a skilled operator. At the longer ranges the time of flight was quite long, and if the operator was subjected to counterfire he might not be able to complete the rather exacting task of tracking both missile and target throughout the full trajectory.

In the 1980s, we are seeing a second generation of ATGMs, using 'semi-automatic command to line-of-sight,' for which the operator has only to follow the target, with the missile automatically tracking along his guidance beam. Here the operator's task was less exacting, but he still had to carry it out and he might be prevented from doing that by some sort of countermeasure, such as being shot at or having the target disappear.

By the 1990s, there will be a third generation, 'automatic command to line-of-sight,' which will require the operator simply to choose his target, point it out and launch the missile, but subsequent homing will be fully automatic. Having recognized the target, the missile will steer all the way without further intervention.

ATGMs can be rocket-propelled from the ground, from a land vehicle, from a fixed-wing aircraft, or from a helicopter, and probably in the future from a remotely piloted vehicle. Or they can take the form of shells fired by artillery and steered aerodynamically during the latter part of their trajectory. They can be guided to home on a spot positioned by a laser designator, directed by an observer on the ground, in a vehicle, in an aircraft, or a helicopter.

By the 1990s, it may be possible to launch very clever submunitions from guns, rockets, or aircraft which are able to seek and find tanks or other vehicles using millimetre-wave radar, infrared, or optical means, attacking them from above or the side rather than through their thick frontal armour.

In general, guided missiles will not possess great kinetic energy, and will need to use the kill mechanism of a shaped charge or a squash head bursting on impact with the tank if they are to hole thick armour or to drive fragments into the interior. Against this type of attack there are new types of armour, using spaced layers of composite materials. The contest will go on between a more effective warhead and a more effective defence.

The other mechanism to kill tanks is to generate great kinetic energy, by the use of a long heavy solid armour-piercing round hitting the armour at high velocity. This requires a gun, whether in a tank, in a tank destroyer, or on its own wheels. The high velocity provides a shorter time of flight than a rocket. At short range it will not be necessary to compensate for the motion of the target, and accuracy will be good. At longer ranges, where target motion matters, accurate fire control will require measurement of range, and computation of deflection, and air resistance will rob the round of some of its kinetic energy.

Thus, it can be concluded that guns will be more effective than guided missiles for the frontal attack of well-armoured tanks at short range. But at longer ranges, and when attack from above is possible, ATGMs of various degrees of cleverness will be more effective than guns.

Land Mines

Land mines can be formidable obstacles to the movement of tanks, other vehicles, and personnel on foot. They are often more helpful to the defence through delay than through infliction of casualties. They are being made more effective than ever before in three ways.

The earlier land mines were buried, and were triggered by the physical presence of a vehicle track or wheel, or a human foot. Obviously the target had to pass directly above the mine. Now it is possible to use remote sensing devices to trigger the mine, and to have it increase its effective range by jumping several feet in the air before exploding, so that it can distribute its fragments over a much bigger area.

The ability of detectors to locate buried mines has depended on the presence of metal. It is now possible to construct mines out of plastic material which will not activate the detectors.

When defensive positions are prepared before the enemy is close, it is practical to bury mines by hand, or with the aid of special digging machines. However, the ability to emplace a minefield quickly in the middle of mobile operations can create an unexpected obstacle to enemy advance and exploitation of success. This is now becoming possible, with dispersal being effected by aircraft, rockets, or guns. The mines would not be buried, but if scattered on ground covered by vegetation they would be hard to spot, especially from vehicles.

While mines can be swept, by hand, by special vehicles, and by explosive devices, the process is slow and dangerous. Unless the attackers accept considerable delays they are likely to suffer casualties from remotely delivered minefields.

Improved Anti-personnel Weapons

There are also improvements in anti-personnel weapons. Small arms are becoming lighter, firing smaller calibre bullets with very high velocity and a high rate of fire. Although these weapons do not have the range of the older types, they have great wounding power at short range. Experience in past wars has shown that there is very little aimed fire by small arms at the longer ranges, so that it is more efficient to provide most of the riflemen with weapons optimized for effectiveness at short range.

Larger munitions launched from aircraft, guns, and helicopters will contain sub-munitions capable of lethal wide area dispersal. This will cause the effects to be more uniformly distributed, instead of having an 'overkill' in a small area. There is also the means of using fuel/air explosives that will spread a blast wave over a big area and direct it downwards so it is much more difficult to get protection by digging in.

There will be many types of booby traps. These self-actuated and delayed-action devices will add psychological and nuisance factors to the casualties inflicted.

Dominant Aspects of the Battlefield of the 1990s

The Importance of Combined Arms Tactics

The first lesson, and we have already heard about this today, is the importance of combined arms tactics. Some of the new weapons just described are likely to be so effective against their intended targets as to neutralize them completely, unless countermeasures are taken. In many cases the most effective countermeasures will make use of systems other than those targeted by the new weapons. The appropriate reaction will be to use combined arms, exploiting the strengths of each to compensate for the weaknesses of the others.

An example of the need for combined arms tactics was offered by the success of Egyptian ATGMs against Israeli tanks in 1973, when these operated without artillery, infantry, or air support. In the same war, when the Israeli air force attempted to operate above Egyptian ground forces equipped with effective SAM and AA batteries, they suffered heavy losses until they devoted the efforts of both air and ground forces to the suppression of those AA defences.

On the battlefield of the 1990s, attacking armour must advance with the co-ordinated support of artillery, infantry, and air, which can exploit the vulnerability of ATGM crews, especially during the flight of their missiles. In addition the tanks will need the support of mobile AA defences to counter the anti-armour capabilities of enemy aircraft, both fixed wing and rotary wing. Conversely, considering the problems of defenders against armoured attack, their ATGM crews will need the support of infantry, artillery, and close support aircraft, and their close support aircraft will need to have the enemy air defence weapons suppressed, whether by air weapons, ground-delivered weapons, or electronic warfare.

From Dr Charters and others we heard about the need for co-ordinated efforts of all arms. I would underline that for the battlefield of the 1990s.

Electronic Warfare

Much of the technology which provides for the effectiveness of the new systems is based on electronics, especially on the electromagnetic transmissions of radar, of guidance signals, and of communications. All electromagnetic transmissions are subject to countermeasures, against which there are counter-countermeasures. It is a highly technical contest, in which success is very dependent on detailed knowledge of the opponent's equipment and up-to-the-minute tactics. Rapid changes in the situation are to be expected, perhaps even day to day, as responses are made on both sides. So field intelligence and technical intelligence will be absolutely key participants in this electronic battle.

Against radar, we can expect jamming and chaff. Against radio frequency guidance signals there will be alerting devices to warn the intended victim, and jamming and deceptive transmissions to foil the guidance. Against laser, infrared, and optical signals we will see smoke, deceptive flares, camouflage, and the use of whatever concealment may be offered by nature. We must expect jamming to be used against our radio communications. With all of these transmissions we can expect detectors and direction-finders to be used to locate the source for subsequent attack, or even for instant attack by missiles homing on the transmitter.

It is possible that instead of interfering with wireless communications, the enemy may prefer to monitor them and obtain valuable intelligence. However, counter-countermeasures are available to make the communications secure.

Because of the threat from anti-aircraft missiles, all tactical aircraft will need some self-protection electronic warfare equipment. Beyond this, there will be a growing need for specialized aircraft to provide EW support for other aircraft attacking over heavily defended areas.

The proper response to received electromagnetic signals may be to radiate a particular jamming transmission, to undertake an evasive manoeuvre, or to launch an appropriate weapon to destroy the weapon that's coming after you. In order to react quickly, making full use of known information, there will be increasing dependence on computer decisions, based on stored intelligence data and a prearranged programme. In many of the latest tactical aircraft it has been judged that computers are faster and better able to cope with simultaneous requirements than a second crew member.

Rapid Movement, Rapid Expenditure of Ammunition

There are a couple of other things that I think will be features of the battlefield of the 1990s. Things are going to move very quickly and ammunition and other technical devices are going to be used up very quickly. In the not very highly sophisticated battle of the Falklands, the artillery nearly ran out of ammunition. In the future battlefield, mobile anti-aircraft are going to move up with the armour, to enable an air defence umbrella to be kept above the advancing spearheads. The newer infantry combat vehicles will provide increased mobility and firepower to the infantry, including an element of organic anti-tank and AA armament. Night vision equipment should allow operations to continue twenty-four hours a day.

The concentrated hitting power of the new weapons, combined with high mobility of modern ground forces, offers great possibilities for the attacking force to achieve a breakthrough followed by deep and broad exploitation. However, the improved capability for the defender to put obstacles into place quickly, and to cover them with powerful weapons may enable him to seal off a breakthrough, and the same potential for firepower and mobility which helped the initial attack could help the counterattack.

In these circumstances ammunition will be expended at a prodigious rate. Since they are very expensive, stocks are unlikely to be large, and those stationed forward will probably be used up very quickly. The whole battle may hang on resupply, on which side first ran out of new weapons.

Significance for Deterrence

It's worth mentioning just one word about the broader picture of the battlefield of the 1990s, which I have assumed to be conventional only.

Whereas new weapons deployed by one side can probably be countered if the opponent provides himself with the appropriate offsetting weapons and tactics, it will be necessary to have these available in sufficient numbers, and to have the commanders and troops properly trained in their employment.

If the Warsaw Pact equips and trains their forces for the battlefield of the 1990s, but NATO remains with the weapons and tactics of the 1980s, it is most improbable that a conventional attack from the East could be contained by conventional means. Deterrence would come to depend on tactical nuclear weapons even more than it does today, and the nuclear threshold would be lowered, a development that would appear unwelcome to almost everybody.

Our alternative is to equip ourselves with modern conventional equipment and to maintain a certain level of conventional capability, or we will depend even more on nuclear weapons and I don't know anybody who wants that to happen.

Brian MacDonald

We will have the opportunity for comments or questions.

LCol Larry Motiuk

Dr Lindsey: You have left out nuclear weapons for a good and sufficient reason. What about chemical and bacteriological weapons?

Lindsey

I never know whether they are 'conventional' or 'unconventional.' But I would say that chemical weapons deserve a great deal of attention and concern.

The moralists seem to feel that it is bad to know anything about chemical warfare. I think that not only has the West lost any momentum it might have had over offensive chemical weapons but has devoted far too little attention to defensive chemical warfare. There is no doubt at all that the Warsaw Pact are very good at it, and very well trained in it. They consider it an integral part of military equipment and would probably have no hesitation whatever about employing it. The West has allowed its expertise in chemical warfare to slip at a dangerous, possibly a fatal, rate in the last decade.

One of the steps to rebuild NATO's offensive capability has been the campaign to equip with 'binary munitions' which are much safer to store than the old-fashioned ones. A binary chemical weapon is not dangerous until the moment it is fired. Defensive CW requires a great deal of equipment that is not being deployed widely enough. Realistic training is difficult to do and we don't do enough of that.

Biological warfare seems less likely to be an effective military weapon. If you start making one part of a continent sick it is likely to spread and everyone will catch whatever it is. Also, the effects are likely to be rather slow and unpredictable in their extent. The advantage of chemical weapons is that you can arrange that a piece of territory becomes uninhabitable for 24 hours, or as long as you like. But once you start biological warfare, it is hard to stop.

There are attempts to have arms control conventions for both of these, but they are very difficult to monitor. Biological and chemical weapons are very much like medicines that are legitimately made all over the world. It would be difficult to be sure that any agreement not to manufacture them was being honoured.

I don't want to go into chains and things. They are horribly effective unless you have the right defences, and we haven't been working at it.

Hugh Hanson

When I listened to your description of the battlefield I visualized open countryside. To what extent do the kinds of weapons that you describe have a utility in urban warfare? I don't mean guerrilla warfare, but the Stalingrad type of situation, which I don't think is unlikely.

Lindsey

I don't think it is unlikely either. One reason that I restricted myself in the beginning to an open battlefield of manoeuvre was so I could tell you about all the things these weapons can do. A lot of them cannot do it in the middle of a city.

You are exactly right. Long ranges imply a long unimpeded line of sight and you won't get that in a city. A city helps the defender, especially if the defender doesn't mind having the city smashed up. If he wants to enforce delay, he would probably prefer to have the action in a city. If that is true, it may be that the attacker will avoid them. But in western Europe, in places like the Ruhr Valley, you can hardly avoid urban areas; they are beginning to occupy a very large fraction of the whole landscape. I think you could go through the list of weapons and say about each one that it will not work as well in urban conditions.

An attacker who wants to march through a city might use chemical weapons. Quite possibly he would arrange that everybody in the city was dead in a few minutes. It becomes more and more unsavoury as one goes into it.

LCol Bill Little

I recall that at the end of World War II there was a tremendous amount of research carried out in order to arrive at what distances were involved in the sightings of tank to tank. This investigation was carried out again in the early 1960s in northwest Europe on a very large scale. The results were identical in that tank-to-tank sightings in that area were between 800 and 1000 yards. I don't see that changing in the projected battlefield in northwest Europe. Are we going to really require these tremendous velocities, reaching out incredible distances, when the human being cannot see that far?

Lindsey

There has been a great deal of work of the type you mentioned and the conclusions are about what you said, but not always. The ranges may be usually 800 to 1000 metres, but there are cases of 2000, 3000 or 4000 metres. The intervisibility depends very much on the type of ground. The western desert gave long ranges. Mountains might give you long in some areas and not in others, and in the woods it is very short indeed.

If the attacker knows that there are long ranges open to the defender, he may render them unusable by smoke; or he may put smoke ahead and then appear at the last minute. I think it highly doubtful that the clever long-range weapons will get very many chances to be used. But if you can ever catch your enemy under conditions where you out-range him, that is very good. In the Falklands, incidentally, the British artillery out-ranged the Argentines by quite a lot and that helped on a number of occasions.

Probably a lot of these clever things will only be able to be exploited infrequently and obviously some sort of short-range highly reliable weapon has to be provided as well. There is a certain parallel with aircraft. Perhaps too much attention is being paid to great capabilities that will hardly ever be realizable. Moreover, very often it is foggy and your visibility is cut in half. That is a very severe limitation.

HARRY HALTON

HIGH TECH DEFENCE PRODUCTION AND CANADA'S INDUSTRIAL BASE

Brian MacDonald

Ladies and gentlemen: At this stage of the seminar we move from a consideration of the technological environment to a consideration of the technological resources that we have to cope with that environment. Colonel Stevenson will introduce our next speaker.

LCol Michael Stevenson

Ladies and gentlemen: Mr Harry Halton is the executive vice-president of Canadair. He was born in Czechoslovakia and educated in England. He has participated in the engineering and development aspects of all the Canadair projects. As programme manager, he directed the development of the Canadair surveillance drone, the Canadair CL-215 water bomber and the Canadair Challenger business jet projects. He was appointed director of engineering in 1970, vice-president of engineering in 1972, and to his present position of executive vice-president in 1975.

Mr Halton is a Fellow of the Canadian Aeronautic Space Institute, a senior member of the Instrument Society of America, a member of the Institute of Astronautics and the Institute of Electrical and Electronic Engineers. He is currently the vice-president of the Canadian Aeronautics and Space Institute, and a member of the board and the executive of the Air Industries Association of Canada. In 1978, *Aviation Week* named him as the only Canadian among the 15 who have contributed most to the advancement of aeronautics. In 1980, the Governor General of Canada presented him personally with the Organization of Rehabilitation and Training Centennial Medal. In 1981, the Canadian Air Force Association awarded him the Gordon R. McGregor Memorial trophy in recognition of his contribution to Canadian civil and military aeronautics.

Harry Halton

Thank you for that marvellous introduction - mostly undeserved, I might say.

Ladies and gentlemen: I have set for myself this afternoon an ambitious objective. In the next half hour I plan to describe to you a capability and capacity which we in our Canadian aerospace industry have built up to support - in terms of equipment and logistics - the Canadian Armed Forces as well as those of our NATO and other allies. This is an industry with annual sales of almost $3 billion, of which about 75 per cent represents export sales, which contributes significantly to our country's balance of trade. The industry spends $250 million a year on research and development to keep it on the leading edge of technology.

I am not going to be modest about the industry's first-rate achievement. But, before I begin, there should be a word of explanation.

Aerospace is an industry where often it is difficult to differentiate between strictly military and strictly civil product applications. Indeed, it can be said without fear of much contradiction that the technologies on which most aerospace products are based, around the world and to a large extent in Canada, were generated by military programmes and funded out of military budgets. There are exceptions, of course - but these are few and far between. There are other reasons, too, for this inability to differentiate between military and civil applications. Some of these relate to the fact that the environment in which the aerospace industry's products are operating is common to the military and civil. The military and civil operators, after all, use the same airspace, use the same airwaves, and operate into the same world-wide geography. Other considerations such as safety, reliability and maintainability also are common to all aerospace products, be they civil or military. The differences, such as they might be, are more in quantitative rather than qualitative terms.

All of this then tends to make differentiation between the military and civil sectors of the aerospace industry difficult, if not impossible.

In its product line too, the aerospace industry very often caters with the same basic product to the military, paramilitary and civil requirement. A civil airline type aircraft – often with very minor modifications – will become a most efficient military transport aircraft. A civil transport with the addition of specialized equipment and by the introduction of certain performance modifications, can become a very efficient maritime surveillance and ASW airplane. There is a high degree of commonality between cockpit instrumentation of military and civil aircraft, as there is in many of the systems, hydraulics, electrics, pneumatics, environmental controls, and so on.

In the avionics industry, most especially, the commonality of products is very prevalent. Only those items of hardware which are specifically associated with the offensive or defensive missions of the equipment are peculiarly military.

With this introduction, then, let me proceed with a description of the industry. Sales have grown in a decade by more than 400 per cent. With annual sales close to $3 billion, it is in the forefront of the economic spectrum of high technology industry.

The industry also represents a very important source of export revenue to Canada. In the last decade, more than 75 per cent of its product was exported. Indeed, if it was not for its export trade, the aerospace industry could not sustain itself and would not be able to provide any reasonable level of support to the Canadian Armed Forces.

In employment too, the recent downturn notwithstanding, the industry has shown steady growth and we would expect that when a normal economic environment is restored, the industry will likely level off at perhaps 45 000 employees.

Canada holds a position of world leadership in the development and production of small turboprop, turboshaft, turbojet and turbofan engines in the 500 through 3000 SHP class by Pratt & Whitney of Canada in Longueuil, Quebec. The PT-6 Turboprop powers an innumerable number of aircraft and aircraft types in Canada and around the world. The PT-6 Twin-Pac powers a great many helicopters of the world's rotary wing fleet, and the JT-15 Turbofan is foremost in the class of small turbofans, powering as it does the popular Cessna Citation aircraft. About to enter service on several aircraft types is the new generation of PW-100 Turboprop engines.

Canada, as exemplified again last week, is also highly active in space-related activities, both as a supplier of satellites as well as systems and sub-systems. Spar Aerospace in Toronto and Montreal are responsible for the Anik series of communications satellites (one of which has just been placed in orbit during last week's flight of the Columbia space shuttle) and the development and production of the remote manipulator systems colloquially known and admired around the world as 'Canadarm,' a model of which was also on board the Columbia space shuttle.

Supporting these prime manufacturers as well as developing their own products for export to other countries, is a whole spectrum of system and sub-system manufacturers. Some are household names in aerospace around the world and for the sake of brevity, I shall mention only a few and an even fewer selection of their broad range of products: Computing Devices in Ottawa with such equipment as the projected map systems used on a variety of military aircraft; Garrett Manufacturing, here in Toronto, are the predominant supplier of electronic environmental controls for a great number of civil and military aircraft; Leigh Instruments in Ottawa, with such devices as emergency locator transmitters; Litton, here in Toronto, best known for their inertial navigation systems; and Canadian Marconi, in Montreal, with such products as cockpit display systems.

Last, but not to be forgotten, there is a whole chain - across the country from the Maritimes to British Columbia - of processing sub-contractors providing support with special materials and processes, castings, forgings, machine shops, and many other products and services. Many of these, because of the aerospace industry's special requirements for quality, reliability, documentation and source control, are specifically dedicated to aerospace.

Let me now turn to the aircraft prime manufacturers. DeHavilland have supplied, from their Downsview plant, what has come to be the real workhorse of the flying community, many hundreds of the DHC-6 Twin Otter. The DHC-5 Buffalo is in the military transport inventory of several lands and the Dash-7 has become a most popular commuter airliner with many regional and local airlines around the world. DeHavilland is now configuring the Dash-7 as a potential maritime patrol and surveillance aircraft for air force and navy use. Under development and soon to fly is DeHavilland's Dash-8, an aircraft which - being larger than the Twin Otter but smaller than the Dash-7 - will surely find its way into the inventories of the civil and military fleets around the world.

Canadair in Montreal is building the CL-215 amphibian, originally developed as a specialty forest-fire fighting aircraft, has since been configured both for military search and rescue operations as well as military personnel and logistics transport. Awkward and unbecoming as it might look, it is nevertheless a highly efficient piece of machinery in its many civil and military roles. A real testimonial to its capability and quality is the fact that nearly all customers have reordered aircraft following their initial buys.

The Canadair Challenger now in service with private as well as government operators is an example of all that is best in Canadian advanced aerospace technology: the use of advanced transonic aerodynamics, the use of advanced metallic and non-metallic materials, the use of the most modern system and subsystem technology and the application of advanced structural ground and flight test and development techniques. The aircraft is now being considered for various specialty missions in military and para-military roles.

I will now spend a little time on battlefield surveillance. This is, I believe, most appropriate following the previous speaker's remarks on the high technology battlefield of the 1990s. I might say, incidentally, that for this purpose I am using the same definition of the battlefield as Dr Lindsey used earlier. Little known but nevertheless true, is the fact that we, in Canada, and specifically at Canadair, are the world's leader in unmanned battlefield surveillance systems of the penetration type. Penetration, in this respect, means forward of the forward edge of the battlefield.

For the past more than 20 years, we have been developing, producing for several NATO countries and supporting what was, and still is, the most advanced operational system of its kind. The CL-89 (in NATO parlance, AN/USD/501) drone is the result of a series of operations research studies in the 1960s of the battlefield as it was envisaged for the 1970s and 1980s and the related requirements of the artillery commander for intelligence. These operations research studies established clearly that there was a requirement for a penetrating reconnaissance and target acquisition vehicle for the division commander's area of influence which had to provide high-quality intelligence but with reaction and response times significantly better than the reaction and response times characteristic of air force operations. Furthermore, such a system would need to be capable of being organically integrated within the artillery structure and be, generally, compatible with army skills and logistics capabilities.

The battlefield intelligence data, apart from being timely, had to be of a kind which lend themselves to the type of interpretation which the battlefield commander needed in order to perform his mission. In addition, the penetrating vehicle had to be survivable in a battlefield environment and against a threat of the 1970s and 1980s.

Simultaneously, in the early 1960s, a Canadian operational requirement, followed by a British general staff target, subsequently ratified by a STANO operational requirement, gave credence to the results of these operations research studies performed by Canadair.

These operations research studies were followed by many systems analysis and trade-off studies. From these emanated a system which would launch from a zero-length launcher by means of a solid-propellant booster, a fully preprogrammed drone powered by a turbojet sustainer engine, carrying a photographic camera as a payload and which would return to a designated recovery area, decelerated by a parachute and landed by means of a bag-cushioning system, so as to be fully reusable for another flight. The system was to be fully mobile, using specially adapted but standard army vehicles.

Canadair then proceeded with the development of such a system starting in 1962 and the CL-89 was born, and with it a brand new business line for Canadair and Canada. Since then, Canadair has produced more than 600 drones for five NATO countries who have flown more than 2000 fully operational training flights and many tens of thousands of simulated flights.

Many improvements to the system, including the option of carrying an infrared linescan sensor, the ability to fly several flight altitudes on the same flight, automated mission planning and the ability to fly some very complex flight paths were added. All of these are the results of continuing operations research and analysis efforts.

As a result of the success of the CL-89 and the recognition of the greater sophistication of the battlefield of the 1980s and beyond, and acknowledging the longer range of artillery weapons as well as the new tactics associated with the activities deeper behind the forward line of the troops, a requirement was established for a new more advanced system which would provide intelligence to the corps commander in his area of influence. Thus, the CL-289 activities were initiated in 1977 and the system, having completed its development, is presently undergoing service trials at the proving ground in Yuma, Arizona.

This system, though bearing many similarities to the CL-89, is much more complex, has much longer range, provides simultaneously photographic as well as infrared imagery, some of which can be obtained on the ground in near real time, as well as providing a nearly infinite flexibility in flight patterns, including a quasi terrain-following capability.

With this system, we in Canada and we at Canadair, this time in co-operation with our associate contractor, Dornier of Friedrichshafen, Germany, are maintaining the momentum and our leadership in this field of endeavour.

We at Canadair are not doing it alone. Other companies in Canada are contributing in no small way. I shall quote just two of many examples. Bristol Aerospace in Winnipeg designed, developed and are producing the solid propellant booster motor. Such a development in itself is no small task. The requirements are most stringent. For example, the thrust decay at the end of the burn must be most carefully controlled in order to ensure a satisfactory separation. Irvin Industries in Fort Erie have designed, developed and are producing the parachute and landing-bag systems. Again, not only are the requirements most stringent, but are in many respects unique and are pushing current state-of-the-art.

Quite early during our operations research activities, we recognized that a shorter range system for the brigade commander's area of influence would be required. The mission would be reconnaissance, target acquisition, fire correction and damage assessment. These studies indicated that, for the high technology battlefield, a true remotely piloted vehicle, capable of both hovering as well as translating at reasonable speeds and carrying sensors capable of providing intelligence in real time, would most likely satisfy such a requirement.

Fortunately, we at Canadair, apart from our surveillance technology, also had the technological base related to VTOL rigid rotor technology as it emanated from our CL-84 tilt-wing VSTOL experimental project and this led us to the design of the CL-227, commonly known as the Peanut.

This air vehicle, completely modular in concept, consists of a turbo-shaft power plant driving contra-rotating collective and cyclic-pitch rigid rotors and carrying a TV camera. It launches from a simple take-off and landing platform. It is controlled by an operator in the control room while another operator is able to interpret imagery in real time.

We have recently completed a series of full-scale development and free-flight tests and are now planning for more development and demonstration testing.

We have been able to generate what we consider unprecedented interest in the surveillance and battlefield intelligence gathering community in several countries, but also with different payloads for other military and, eventually, hopefully, civil applications.

It is perhaps now propitious to return to the aerospace industry as a whole. All of this activity and all of the products that I have described to you don't just happen. Industry spends close to 10 per cent of sales on research and development.

Who is performing this $250 million of research? It is a staff of 4240 of whom approximately one half are professional engineers and scientists and the other half are technicians. Approximately one third of the total staff has Bachelor's degrees.

In addition, and very important, the industry enjoys excellent support from R&D laboratories in government and the universities. These facilities are very much part of the total aerospace industries' R&D complex and without them, the industry could not do what it does or be what it is today.

So let me conclude by saying that we in the aerospace industry are particularly proud of our achievements in Canada. The products which we have developed are second to none in the world's markets. The capabilities which we have produced not only operate economically, productively and competitively, but provide an excellent preparedness base for Canada's future defence requirements.

But let me add a word of caution. To a large extent, our present aerospace industrial complex was established during the 1950s and early 1960s – during the period when large domestic military procurement provided much of the technological impetus to Canadian industry. In recent years, much of Canada's major procurement has been so-called 'off-the-shelf' from abroad. Such procurements offer little opportunity for an involvement in the generation of technology and the advancement that is inherent in such programmes.

An industry such as ours needs continuous nurturing by the injection of new technology. This does not happen by itself. It requires continuous and persistent attention by the Department of National Defence and industry. That attention is our task and - in the face of continuous technological change - that is the challenge.

LCol MacDonald

Thank you very much, Mr Halton. I will now take questions.

Dr W. A. B. Douglas

Mr Halton, I am writing volume two of the official history of the Air Force. During the Second World War, we had an enormous aircraft industry and a great training plan but had no control over the resources. We built largely for export because the balance of payments required we build more for export than for ourselves and we were always at the bottom of the list of priorities for good equipment and good aircraft.

Do you foresee that we are in a position to prevent such a thing happening again?

Halton

I believe the answer is that we hit bottom some time ago, and are now back on the way up. The question is not whether we are going up, but the gradient at which we are going up. It is a very, very shallow gradient.

MacDonald

Has the aerospace industry been able to take advantage of the offsets with respect to the recent major equipment purchases such as the Auroras and the F-18s and the agreements for offsetting purchases in Canada to any great extent? More importantly, if it has, what advantages has it been able to take of technology transfers or contracts relating to R & D?

Halton

The answer to your first question is yes. The aerospace industry did take advantage of these offset programmes but did not do as well as we might have hoped. It probably did better than it might have done under other circumstances. So it hasn't been bad.

I think it is fair to say that on the CP-140 Aurora programme there has been a degree of technology transfer, perhaps more so in the avionics industry than in the air frame industry. You have to consider that the air frame of that aircraft is an old design. There is nothing much to be learned from the air frame. But in the avionics part of the industry, there has been a large amount of transfer.

There has been no fall-out in terms of R & D projects.

As to the F-18 project, it is too early to tell. We are just at the beginning of it. The aerospace industry has received some programmes, but it is too soon to say how far they will go, how long they will last, or what transfer they will generate.

Dr George Lindsey, Department of National Defence

This morning we heard about the great exploits of the VSTOL Harrier, and this afternoon I said that I think VSTOL is going to be very important in the future. You showed us the CL-84, a tilt-wing with propellers, hoisting itself straight up into the air. Can you tell us whether you think the future of these VSTOL is going to be directed jet or is it going to be like your bird? And what happened to your bird?

Halton

The CL-84 was a research rather than a production project. It was stopped in 1972, essentially when the last potential customer went out of business. That was Admiral Zumwald with his sea control ship concept. The CL-84 was judged to be about the ideal vehicle for the sea control ship, for a small carrier with about a 500 foot deck; a 15 000 ton carrier. Its real advantage was a short take-off/vertical landing capability, because with a 200-foot deck run you can double the payload. It is not a terribly efficient vehicle for vertical take-offs. It really is a *short* take-off/vertical landing vehicle and therefore it has a specialty mission. The sea control ship was just such a mission, and we haven't really found a mission since that really suits that vehicle. It was a research programme which did not, in the end, have a defined mission.

That's where we sit today. There is now in the United States a tilt-rotor programme going as a joint project between Bell and Boeing. I believe that the tilt-wing concept is a better concept than the tilt-rotor concept. I think it will go one day, but will likely be a more advanced prop-fan project, because the prop-fan is probably the right kind of power plant combination for that kind of project. The prop-fan isn't quite there yet, but when it is, I think the tilt-wing programme is going to come into its own.

Dr D.A. Charters

We heard a lot this morning about the need for airborne early warning for fleet protection. Is there a capacity within the Canadian aerospace industry for developing this kind of technology, either on a larger scale or from the small drones you have been demonstrating to us?

Halton

The 227 Peanut type of vehicle is ideal for providing permanent surveillance somewhere ahead of the fleet, and is, in fact, being considered for just such a role.

MacDonald

How far in front of the fleet could that particular vehicle be deployed?

Halton

We are talking of a vehicle that flies at about 70 to 80 knots and can remain airborne for two or three hours at a time. But the control requires line-of-sight, and further it depends on how high you want to fly it in order to provide the kind of resolution that you need for whatever target you are seeking.

Col Terry Liston, National Defence Headquarters

Would you give us your views on the nationalization of defence production within the Alliance? Your company seems to be producing high-technology equipment to meet needs that we do not have in the Canadian forces. Your customers, for example, for your surveillance equipment, are people who are operating at the corps level. Yet in Canada, we are barely re-establishing the ability to produce a rifle barrel and we are unable to produce light machine guns.

There are also the questions of the cost of producing military equipment within NATO, and the competition within countries and between countries to do so. The British have argued that countries should specialize in the development of equipment. What is your view of the state of the industrial base within NATO and how do you think it should develop?

Halton

Maybe the CL-89 drone is a case in point. It started back in 1960 as a Canadian project. About 1962, the British joined in and about 1963 or 1964, the

Germans joined. Guess who was the first country to drop out? It was at the time of unification of forces and the establishment of Mobile Command. The Canadian requirement, which had been the first, was essentially scrubbed. Though there was no Canadian requirement it was highly successful and became a 100 per cent export programme. The CL-289, the bigger one for the corps level, continued because we had the design capability and the Germans understood that and are utilizing it.

Let me take it further. If, in the last twenty years or so, the aerospace industry had been solely dependent on the procurement of the Department of National Defence, that industry would not exist. It is almost entirely export-dependent. Consequently, all the products that it builds must be exportable and therefore there must be a degree of specialization. If you can't buy rifles in Canada, that presumably is because that industry would have to export those rifles and presumably cannot, because everybody else builds them. We always have to find some unique niche for ourselves in order to stay in business.

NIAG has for years tried to encourage this trend towards specialization and has always run into major obstacles because social-economic problems within the country itself have dictated what each country does, although this is less so in Canada than most other places.

General Bell

What is the state of the air industry in terms of industrial preparedness planning? Is there an indigenous mobilization capability for the nation? Is there effective linkage now, not only at the ministry level, to define what the industry role should be in an emergency?

Halton

I am not an expert on what is happening within government. But the aerospace industry in Canada provides, I believe, an excellent preparedness base. The question of mobilization really relates to two things. First, how much time you have to mobilize, and second, what you want that industry to build.

For instance, the requirement might be for the industry to build more, or perhaps more sophisticated items of the kind that it is building already. The industry is well prepared and can do it, not at a moment's notice but at something pretty close to it. If it is a matter of building different things – high technology but different from what it does now – then it is a question of what that is. Is it something that the industry needs to develop? Or can industry get it from somewhere? Either way, it is a time-consuming process.

There is no simple answer. But in general terms, the industry is well equipped, well manned and highly productive.

Major L. R. Sprowl, Trenton

Canadair has built some fine aircraft such as the Yukon, a good workhorse, in which the cost per ton mile was extremely low for an aircraft of that vintage. There is a great need around the world for a good strategic airlifter of medium size with a fairly long range. The C-141 Hercules has fairly poor economics; its specific air range or fuel consumption is poor. Is Canadair thinking of a replacement for the many countries that are flying the Hercules? Are they looking at an airlifter that could have a super-critical wing, a long range, and something like the CFM-56 for the power plant?

Halton

We would like to but it is not very likely. The development of such an aircraft would be somewhere between two and three billion dollars. And that pays for nothing but straight development. You would have no hardware except maybe four or five prototypes. There isn't much likelihood of the Canadian government funding that kind of a project. And it can't be a group of other small countries funding that kind of programme. The United States is probably the only one that could raise that kind of money, but there is not much motivation on the part of the United States government to want to come in as a partner. That's why I say it is not likely.

LCol MacDonald

Mr Halton, thank you very much for your presentation. It is very useful for us to have a senior representative of the Canadian aerospace industry to speak to us about the capacity of that industry and the contribution that it has made to export trade and therefore to the balance of payments of this country.

OPEN FORUM

LCol Larry Motiuk

I would like to get into the area of intelligence errors. I understand there was quite a misreading on the part of both the Argentines and the British at both the strategic level and the tactical level. With the Exocet missile, I guess somebody forgot to tell the British that the Dassault team were in Argentina working on it. Can you comment on the lessons to be learned in the intelligence game?

Elliot

That is very difficult because I don't really know what the intelligence people read on that. There is still a certain amount of attitude in Great Britain that Wogs begin at Calais. There was some comment, certainly in the press, perhaps in official circles, that the Argentines did not have the capability of hooking Exocet to Super Etendard and flying them off. It is a serious mistake in intelligence to underestimate the enemy. Many people forgot that the Argentines do have a fairly advanced industrial base and a pretty good educational system. You may dislike their political structure but it is a mistake to bring that into the technical side.

Col Max Robinson

Dr Lindsey described a wide variety of defensive weapons that could be expected by 1990, including such things as short-range and long-range, and land-based or air-launched weapons. It seems that with our limited resources, we cannot cover all these roles. Are studies in progress to limit our resources to specific roles? Also, which elements of our forces would carry out those roles? Also, what is your view of that type of approach?

Lindsey

I don't think the problem is being handled in the comprehensive, systematic way that you suggest. It would be a good thing if it were. Each environment or service looks at its own requirements. It could be that the navy wants a Harpoon for a naval task and perhaps by good luck somebody else might want something that is almost like a Harpoon for another task. But it doesn't seem to happen that way very often.

I believe that Canadian forces are pathetically weak in the anti-aircraft area. Some of the other speakers have mentioned this. Dr Friedman noted that it wasn't just Canada that is short on anti-aircraft defences. It is quite possible that an anti-aircraft weapon could be developed for multiple roles. But to my knowledge the problem has never been tackled that way. I would be curious to hear what Dr Friedman has to say about that. Remember that we are talking about a tiny base and very small quantities of equipment.

Dr Friedman

I would go back to what Mr Halton said. If your product is good enough, the production base is going to be supplied by export. It is the same problem as was discussed about the transport plane. The price of entry is extremely high because the competition has had a long time to develop. I'm not sure to what extent licensed production helps solve that problem. I suspect that it keeps you employed but wipes out your R&D base. Canada has had in the past some very sophisticated weapon programmes, such as the CF-105, which were wiped out, I believe, by outsiders. I'm not sure to what extent one can arrange co-operative international programmes within NATO, such as a combination of Canada, Britain and Australia. Right now these do not seem to work too well.

Halton

Licensed production is not a very appropriate thing for us in Canada to do. I agree with the policy of the Department of National Defence which is to buy highly sophisticated weaponry off the shelf and arrange for some other means of transferring technology. Some assembly, not licensed production, would be appropriate for Canada.

Friedman

Since the war in the Falklands, with all the publicity given to the attacks on the ships, there may be greater opportunities now for entirely new low-level weapons that could have naval and land applications. If they were ready to be marketed soon, it could conceivably pay off. If you look at small attack boats, 100-foot types, that are very common in the navies of the Third World, you will see they have no air defence. They may represent a valuable market.

BGen C.L. Kirby

What I would like to do is talk about the context of high tech.

Endurance is a desirable characteristic in armies in wars, just as persistence is the necessary characteristic of the regular soldier. If we look at some of the historical applications of high tech, going back to the First War, we find that they appeared in spasms, in applications that were functions of their capabilities, their availability, and their expendability. Endurance comes somewhere near the end of that, and only involves them indirectly. The tank, as an example of high tech in the western desert, evaporated from the battlefield in a matter of days; before its reappearance other lower tech forces held the field. The same thing applied in Russia where, at the battle of Kursk for example, the German armoured forces in enormous strength disappeared in a matter of hours rather than days, and the battle then went back to a lower tech level until more tanks appeared.

In the Middle East, we could look on the various Middle Eastern wars as campaigns stopped for basically the same reason: a rapid expenditure of high tech and then a build-up again over a period of peace until they are applied again.

In the Second World War, it is interesting to note that there were only 14 German divisions with no horse transport. These divisions created the blitzkrieg and got all the attention, while something like 100 or more divisions, that marched on foot with horse transport and horse-drawn artillery, provided the military muscle that kept the campaigns going, and provided the opportunities for the high tech elements to achieve what they did achieve.

Our situation now is similar. Our aim must be to fight a long winning war rather than a short losing war. If we are to do that, we need large forces, many of which are equipped to fight on a low tech mode; or perhaps it would be better to say on as high as possible a mode. There is a strong impression in the Canadian forces (which, I remind you, call themselves "professional" and have done so more and more throughout their decline) that because little Canada cannot afford sufficient high tech to produce only ideally equipped forces, it really can't go to war, or it can't go to war effectively. That may be one of the factors that has prevented this country from having any mobilization plan whatsoever.

We must have a balance between high tech and low tech, and a Canadian doctrine that incorporates both. Those aspects of high tech in war, which are not new, must be capable of use in a long war that we can win and that won't seem to be lost when our limited resources of high tech are expended.

MacDonald

A very cogently argued question. Which one of our panel would like to respond?

Lindsey

I don't think that was a question. I think it was a very good suggestion.

I believe it is quite possible that if a war amongst the most highly sophisticated armies broke out these wonderful things would all be fired off in a few days. Somebody would almost win while the others would just manage to check them. Then they would have used up all their toys and they would revert to whatever was left.

It is true, of course, that NATO has a sort of high-low mix in its force structure for social and economic reasons. The Turks field a large number of very good soldiers but they don't have very many fancy weapons. We might be glad of those numbers. I think it would be presumptuous for Canada to say that we won't go at all unless we have the latest thing. Perhaps we should be strong in the production of something that could be quickly produced in large numbers, perhaps at the low end of the high-low mix. That might save the day some time. It is quite wrong to bet everything on the super toys, only to be left in the lurch without them. That could very well happen, for it is the side who runs out of things first that will lose. If they both run out at the same time, they will revert to a lower level of exchange.

Friedman

There's another aspect. If you have a miracle weapon that doesn't work because there is some catastrophic failure of a trivially simple countermeasure, and you have bet everything on it, it's even worse than running out. The troops crack. I think the mass failure of smart weapons would have that effect. Yet there is a lot of evidence that many of the smart weapons currently in service can be countermeasured relatively easily. You can make a case that a mix is not just protection against a war lasting more than three days, but also protection against a catastrophic error of judgement on one's own part.

Charters

Moreover, high tech equipment tends to be a bit delicate and have serviceability problems. I read at one point that the absolutely fantastic F-14 jet was suffering from about a 60 per cent unserviceability rate. It's a wonderful machine – when it works. If it doesn't work, then you're stuck and you wish you had the simple plane that flies fast and can shoot. Certainly this happened in the Falklands. Very delicate pieces of kit got wet or somebody bumped into them and vital wires and springs broke.

Rod Byers, York University

Can we broaden this discussion to the political aspects? I was struck by problems of first, civil-military relations, and secondly, problems of inter-service rivalry, not only in the United States and the United Kingdom, but also in the Canadian context, given General Kirby's observations. It strikes me that, it is not so much whether arguments will go on within the military, but whether our politicians will learn any lessons from the Falklands. I believe that in the United Kingdom there will be more of the same (that is, Tridents and nuclear weapons), unless there is a change of government. The American evidence of the past 15 years, including Vietnam and the failure to get their hostages out of Tehran, indicates that their politicians have learned nothing. I don't think that politicians in this country are going to learn anything from what has gone on either.

Elliot

As a Canadian living in England, what impressed me most was the fact that this was really a British war. It was not a war for Margaret Thatcher or the Tory party. Right across the board, there was a great deal of support for going out and hammering Argentina. This came as a shock to the pundits (in the *Guardian*) who regard war as a nasty thing that nobody wants.

One of the stories that comes out of the war, which may or may not be true, is that Margaret Thatcher went down to Admiral Fieldhouse's headquarters very early in the crisis and was so impressed by the sheer competence of the place that she said 'What can I do?' And she was told to keep the politicians out of the way, which effectively is what happened. There was very little political interference at the time. Whether or not this would have gone the other way had there been a significant setback or defeat is problematical. Thatcher rallied the political structure around the war effort and went on from there.

The chain of command was very simple: cabinet office to chiefs of staff, chiefs of staff to Fieldhouse, Fieldhouse to task force. It would be normal on the Canadian scene, I suggest, with direct communications, to have everybody and his brother trying to chipmunk on the side. This did not happen. Maybe there is a lesson for us there.

Byers

Our Defence Ministry said that, as a result of the Falklands war, we will have to change some of the capabilities of the Canadian patrol frigate programme since the original design capabilities and the weapons systems would not allow the Canadian patrol frigate to operate in a high threat environment. But are our politicians going to learn anything? Are the politicians in the United Kingdom going to learn anything from the semi-dismantling of the Royal Navy? If the Argentinians had attacked six months later, Britain could not have mounted that operation. Are the politicians learning anything for the future?

MacDonald

Professor Byers is focussing on the periods between wars, in which the politicians must make resource decisions in terms of the funding of defence equipment and defence structures.

Elliot

The British White Paper that was drafted before the Falklands crisis has been issued, unchanged except for a small piece in the middle which indicates that the paper was prepared before the war and there will be further studies. I believe there will be very little change in the direction of that White Paper. There will be minor shifts. I think we will see Phalanx, for example, going into the British major units, if not the minor ones. *Illustrious* has, I think, two if not four Phalanx installations. *Invincible* is getting them. However, the cuts will go on, more or less in the same direction as before. There may be a change of heart over *Fearless* and *Intrepid*, although I notice that *Intrepid* is being decommissioned yet again. They may not sell *Invincible* for another two years. The scrapping of

Hermes may be delayed for two years. But the downward trend that we have seen will go on.

Friedman

I would like to point out a couple of things. First, it was not clear before this happened that the Trident decision made sense. Apparently Mrs Thatcher took it seriously as a matter of national prestige. The whole issue of why Britain should be spending very large sums on nuclear deterrents of her own is a very delicate one in Britain. Telling them that they should go to a conventional type of warfare capability and that the Trident is bankrupting them is very hard for them to swallow. For example, if Labour came back and scrapped Trident it would not be in order to spend more money on real forces. The British have a severe problem in that they have been unable to tell themselves what they need. The Royal Navy is terrible at public relations. It has been totally unsuccessful in making a case for forces. That's why it gets wiped out every time there is a White Paper. That's why the RAF does so well. It's politics, and it has been politics in Britain since about 1920. The Royal Navy has been in trouble for many years.

In the United States, it does not seem likely that anything will be learned from this because we have spent most of our time telling ourselves that we already know it all. There is a 'Lessons Learned Committee' and it is my understanding that they have spent most of their time patting themselves on the back. You would have to scale everything up and then it might look a bit like what we might expect to come against if the Soviets go after carriers, but I don't think that way of looking at it has surfaced at all. The fight over the two new carriers is much too hot, much too political and as far as I can tell, there is very little interest in real analysis. The United States defence scene has become, if anything, more polemic over the last five years than ever before.

It is very depressing. The lesson that politicians should stay out of it should have been learned with the Tehran raid. That was not a high tech issue. That was a question of Jimmy Carter trying to run something from Washington. It was also a question of who takes responsibility. If they had told that colonel, 'You win or you'll be fired or die,' something might have happened. Instead they said, 'You do it or tell us why not.' Of course, there were lots of good reasons why it didn't work. As long as that is their attitude, they will get lousy military performance. It was encouraging that Reagan was not immediately informed when the encounter over the Gulf of Sidra occurred. That's the most encouraging thing I've seen and that's not a joke! The President should not get himself involved in every element. That's something new, but I cannot believe the next President would have the brains to do that.

Charters

The thing that concerns me is that there is a real danger that we in Canada end up preaching to the converted. So far as I can see, there is essentially no defence debate in Canada. Defence is a non-subject. There are not many people in the media who are interested in it. There are only a few with the capacity to analyse it and you get abominable coverage of armed forces and defence issues by people who don't even know the right questions to ask. If that's what we have, you can hardly expect the politicians to be well informed and well advised. Therefore, I don't think our politicians are going to learn very much from this.

Leonard Bertin

I am disappointed that this morning Dr Charters prefaced an excellent review of the subject by knocking the media. At a conference at the Science Centre just two weeks ago of the Royal Society of Canada and the Science Council there was a review of the things that really count in determining what money is spent. The conclusion of this meeting was that it is not briefs from learned people like Dr Lindsey that determine what the politician does. It is what the politicians perceive to be the wants of the great public that will either elect or reject them. They only think three or four years ahead, not the ten or twenty years that are needed to develop new weapons. Therefore we must learn to nurture the media since they are the only contact with the people.

As a separate question I should like to ask who are we really training now to fight for us? I see a great dichotomy between the traditional general, who knows how to command men and get the best out of them, and the new general who will know the right thing to pull from the computer and who knows what the computer will do for him. War ten or fifteen years from now will be completely different at the command and communications level from anything we have been talking about today. It's going to be revolutionized by jumps in the number of things you can get on a chip. Are we going to train a lot of people to make black boxes work – so long as there isn't a high voltage pulse going through because of a nuclear weapon that has exploded somewhere. If we do put great reliance on these people, are we going to forget about something that Professor MacNeill of the University of Chicago talked about very eloquently in his book, *The Pursuit of Power*? According to him, there really is something in square bashing, but unfortunately the people who are best with black boxes do not have any respect either for square bashing or for sergeant-majors or NCOs – or even for generals. They love their black boxes and provided they are getting something out of them, they are happy. Are we then going to train forces who are very good indeed with things that may not work when the time comes? Will they be utterly useless in their alternative functions, in just the same way as those guns on the ships which couldn't traverse once they'd lost power and auxiliary power because nobody could get into the turret because it's not made to be operated by men?

MacDonald

I can see Colonel Liston dancing from foot to foot. That was a marvellous promo for his lead-off address tomorrow morning. Would any of our panelists like to comment?

Lindsey

One of Mr Bertin's remarks about journalists made me want to say something.

I asked one of the Royal Navy people who were briefing us about the Falklands how many research analysts they were able to take. His answer was that they only had 29 spare bunks and there were more than 29 journalists who wanted to go, so no scientists went at all. That is an indication of the priorities between journalism and science as seen from Whitehall.

As for your second point, I think one should separate the problems of commanders from the problems of technicians. Everything you said is really applicable when it comes to technicians - the people who make the black boxes work. But I'm not so sure that the commanders are going to be surrounded by black boxes. I don't know that their choices are going to be so terribly different. If the black boxes don't work, they are not going to succeed, but I don't think the commanders are going to have to be so absolutely technical from morning to night as you suggest. I'm not speaking of the commanders of a one-man aircraft, but of commanders who decide on the major manoeuvres of a big organization.

Bertin

I'm wondering if we will develop a different sort of person who doesn't have the military élan any more?

Friedman

I can give you a couple of examples. The first example is the way navies responded to the tremendous explosion in electronics in World War II. You had traditional naval officers who certainly liked the naval equivalent of square bashing, but they were able to adapt very rapidly to the possibilities of radars and sonars which had been inconceivable when they had been junior officers. When you have a piece of electronics, how it works should be irrelevant. What matters is what it will do for you.

There is a second and cautionary example. When we went to nuclear power in subs Admiral Rickover who ran the programme, was an engineer, not really a combat type. He became obsessed with preventing accidents. To this end, every officer commanding a nuclear ship is a nuclear engineer. But what you really want in a submarine commander is a pirate. You need a man who doesn't care so much about survival. He has to be very aggressive. Before World War II, we had technicians running submarines. They were not very aggressive and we fired about two-thirds of them during the first year of the war.

A third point is that if the chips hold the promise that we claim, you will cut down the number of people you need to maintain them because their reliability is going up. Remember the comment earlier that Standard has no warmup. It just sits in a box and when the time comes - bang - you fire. No technician, no maintenance. That's the goal. Somebody has to know what it can do, but doesn't have to know what's inside. From 1942 until now, those things have been so tricky that the number of technicians has been overwhelming. But we cannot tolerate that forever.

Halton

I'd like to add to this question of reliability. One of the fundamental challenges we have in manufacturing is to make things soldier-proof. We must make equipment that will operate in the same environment as the soldier.

There is a lot of misunderstanding about what high tech represents. High tech represents making the simplest piece of equipment completely soldier-proof so that it cannot fail, so that you don't need the men to maintain it. If the sacrifice in this is that performance may be five or ten per cent lower than the absolute achievable, that sacrifice is well worth it.

Charters

As a response to Mr Bertin, I would dearly like to see a Canadian corps of defence correspondents, but I don't see it happening at the moment. If it doesn't happen, I don't think we can expect politicians to take defence seriously as an issue. At the moment there is a great ignorance of defence issues and broad strategic and national security issues as well as a certain degree of hostility.

George Bell

It has occurred to me, when we look at satellites and lasers and other devices for target acquisition and target control and direction, that the field of command control and communications is becoming highly sophisticated. We are really talking now of a global umbrella whether forces are sea, air or land. What degree of integration will take place in the decade ahead and how many resources are going to be required to achieve that sophistication? I can see that it will affect our need in NATO for space communications and systems.

Lindsey

I think the world-wide communications and the other services that can be provided by space vehicles will have far more effect on the larger allies and on the NATO structure than on the smaller partners like Canada. We have a very serious communications problem across the Atlantic, and these developments are well on the way to solving that for us already. I think that most of the new applications will come at higher formation headquarters where there might be the odd Canadian, but it will not be a Canadian project. Integrating the communications systems of the Western world is very important, but I cannot see Canada having a big stake in it. We have done rather well in space technology, and it may be that we could provide some of the equipment.

I'm not speaking here of battlefield communications where, of course, every unit wants better communications.

Bell

I am thinking in terms not only of communications, but also of satellites, geodetic reference, target acquisition, target direction, and target control right down to the shoulder-carried weapon.

Lindsey

I think the big significance there is in strategic weapons and the direction of carrier battle groups and things that are just too big for us to cope with. I rather doubt that the identification of a target coming over the nearest bush for Private Snooks to aim at is going to come via an earth satellite or from Mr Trudeau's office!

Bell

The second question is for Dr Friedman. I would like to be clear, Dr Friedman, on your thoughts about strategic basing. You implied, of course correctly, that if you have enough bottoms it is possible to move naval forces around the world and sustain them wherever you want on the oceans. What are your views in terms of the whole question of projection of power? When we look at the Persian Gulf, or at maintaining forces in the Indian Ocean and around South Africa, it would seem we need to have Simonstown and Diego Garcia in order to be able to develop and sustain those forces' capabilities. We must be able to project power into the area of a battle at the highest rate possible. When you are at strategic distance, like the Indian Ocean, then maintaining naval forces on station requires battalions of marines - not divisions, not brigades.

Friedman

Diego Garcia and Simonstown are very different issues. How far you are from the battle area determines how fast you can react.

Remember how the media were surprised that the task force took three weeks to get to the Falklands? Everyone reacted as though this was a political decision by Mrs Thatcher to allow the negotiations to continue. No one divided 8 000 miles by 20 knots and found out the real reason! Ships don't move that quickly and ships are the only way you can move anything heavy. The reason that Diego Garcia makes such a difference is that if we want to put anything in the Indian Ocean three days away from where it's happening, then we need to have the stuff stockpiled there. Those forward deployed logistics ships are absolutely essential.

You could operate without Diego Garcia if you were willing to accept the three weeks delay between when the war starts and when you get your forces there. It is really a function of your strategic intelligence and also of your will to operate.

For example, the British clearly had some intelligence that something was going to happen in the Falklands. Let's say that Mrs Thatcher's government had the desire to set it all in motion as soon as anything happened in the South Atlantic - which of course is politically ridiculous. Then presumably the stuff would have been en route as soon as the Argentines tried to seize the island. The British would have said, 'We have 8 000 troops on their way and we hope you will enjoy meeting them.' However, because they didn't have it all en route and didn't have anything floating at Ascension, there was a three-week wait before anything could occur.

If you only want to bomb someone, and you don't care about landing marines, then you can achieve one strike in about 20 hours with B-52s. But if you want to do it on a sustained basis, you require a carrier group - probably at least a week by sea away from any place in the world where such a need could suddenly occur.

Postwar history says that there are a lot of places in the Third World which can explode but that almost invariably we predict them wrongly, usually disastrously wrongly. We Americans (the British are more politically sophisticated) tend not to realize the extent of the disasters that can occur in the Third World, and therefore generally preposition forces some place where we should not be. For example, Diego Garcia does us no great good with respect to Latin America.

Further, there is only so much stuff to pile up. We only produce so many tanks a year. It is quite true that a marine brigade can be overpowered, but we can't put in more unless we are willing to take a lot longer to get there. Basing is not what does it; it is how much stuff you can put in any one base.

If you believe that we will be engaged somewhere in Africa in the next five years, *and* that we can politically afford to deal with the South African government, *and* that the place is far from Diego Garcia (let's say, on the other coast of Africa), then it would pay us to negotiate for Simonstown. Other than that, the best we could hope for is that in an ASW conflict in the South Atlantic, they might give us the right to stop over. But for prepositioning, God help us.

Bell

Are we saying then that in the West, out-of-area operations are only possible on the basis of floating naval bases, the most expensive ones to maintain? Or are we saying that we learned from the Falklands that if we hadn't had Ascension Island, we might have been weeks and weeks away?

Friedman

No. We are saying that the cost of going some place out of area, if you have nothing there already, is a long delay. It costs a lot of money to maintain a force at sea. We have invested a lot of money in a fleet train. We can invest more. Bases in general tend not to be there when you need them.

That's the other end of it. Diego Garcia we virtually own. The British have given it to us. The RDF is based in a lot of other places around the Indian Ocean. The assumption, a bizarre one politically, is that the RDF will be required if Soviets come down into northern Pakistan or northern Iran. Then everybody in the region will line up behind us. But you know, and I know, and the Soviets know that we would go crazy if they did that and that they would run a risk of a really big war. The Russians are not stupid and probably would not do that. So, what is the real contingency? It is some kind of local insurrection where it takes us three weeks to decide if we should go in. And the locals would have the good sense to blow up the local port and airport to keep us out. Do you think that the other countries in the region will be enthusiastic to help us?

Bell

No. I agree. But I also think that our ability to project power is so small at the moment that if that contingency did develop, you would not have the land capability or the air capability to deal with it. The RDF projects a minimum of power at the point of attack. Have we not learned something strategically from this 8000 mile venture that should be examined by NATO? We have been looking at it for ten years and we have done nothing about it. The United States, France and Britain really have no capability in the Indian Ocean.

Friedman

I don't think Ascension played that big a role. It has been overplayed because without Ascension the RAF would not have been operating. But I would say that if the RAF had not been there, the war would have ended roughly the same way.

So they proved that they do have the capability to extemporize out of area, 8000 miles from home, if you can afford to wait three weeks for it to happen. It's a matter of arithmetic and there's no way to get around the fact that it takes three weeks to go that distance. Another point is that a base in the area can be neutralized. Diego Garcia could be neutralized by mining. I don't believe that the bases will save you. The trouble is that you shouldn't have lost the empire! When you had that, you could go anywhere!

Bell

What we are really saying is that the force which is on interior lines has the greater capability.

Friedman

He may have the greater capability now, but I can move more stuff 10 000 miles by sea than they can move down that Ho Chi Minh trail. But I can't start moving it for a few weeks. If the war ends in three days, I've lost. But if I can drag it out, by bombing him with B-52s or showing up with a carrier group, then three weeks later or three months later my superior weight of sea power will tell.

If he can win fast, I lose. It's not exterior versus interior lines. It's time. There are a lot of wars that don't end in three days. That's why it still pays me to work for the navy.

Dr W.A.B. Douglas

I'm going from the sublime to the ridiculous, I'm afraid.

You mentioned naval experience in the Second World War, an excellent example and precedent for our whole high tech experience. I want to issue a cautionary note about the Canadian technical experience at sea in the Second World War. We developed our own radar and we were very proud of it. We have recently produced a book by Knowles Middleton, saying how good we were. But lest anyone be deceived, we had the SW1C and the SW2C which did not work well at all. (I can see Bob Hendy nodding his head.) The type 271 radar produced by the Royal Navy was far better. Then we developed our own centimetric radar and insisted that all our new ships have *our* centimetric radar. It was not sailor-proof. It's even more important to be sailor-proof than soldier-proof. This recent history of radar points out that the only reason it didn't work was because the sailors did not maintain it properly. We must be careful not to get so carried away by our own ability to produce gadgets that we lose track of the need for maintaining them.

Friedman

Amen to that one.

Commander K.J.M. Ayers, Royal Navy

I have a few comments rather than questions.

First, Dr Friedman, on the subject of inter-service rivalry, was absolutely correct in saying that the Royal Navy, the RAF and the Army spend all their time fighting each other for every pound they can get. But the Falklands crisis was one time when all three services worked together and with a real will at all levels. I don't think any service would dispute that.

To prove my point, I will come to the defence of the RAF after the slating they got from Dr Friedman this morning. He said that the Vulcan raids did not achieve very much. But surely it caught the Argentines' attention. When an aircraft can go and drop ten tons of bombs at that range, any Argentine could sit down with a pair of compasses and continue the circle round and see that his whole country was well within range of that Vulcan. Perhaps the Vulcan only dropped one bomb on the runway and did not take the runway out. But it made the Argentines think about the possibilities of further air strikes.

The poor old Hercules got very little mention this morning. I want to point out that those Hercules did a superb job, admittedly staging through Ascension. Some of those aircraft were flying as much as 28 hours non-stop to get the goods down there. The naval forces very much appreciated that.

The navy's air defence is an ongoing question, one we will not be able to answer until all the analysis is complete. But I would question one of Dr Friedman's points about the North Atlantic and the Norwegian Sea and the possibility of the Backfires taking out our ships. He is absolutely correct. They are poorly placed at the moment to fight against something like the Backfire. But what would the Backfire's target be? Would it be a small ship towing some sort of listening device around the ocean? Or would it be the stream of reinforcements going across the Atlantic to make sure that the war in Europe can continue? To my mind, that stream will be one of the main targets, and as soon as the American cavalry appears over the horizon, there is your next target. So perhaps our ships might not be in the front line as far as air defence is concerned.

Finally, I would leave the panel with a hypothetical question, since we are looking at the question of high technology and just how effective the weapons systems are. Can I give you a picture of a finger-four formation of jets coming in low level to do a bombing attack on, say, a group of ships, and they are detected. The first ship fires, say, a Sea Cat missile at the lead aircraft. The lead aircraft breaks hard right into his wing man. He takes off the wing man's wing and he crashes into the sea. The leader recovers and heads back straight over the ship and heads inland. As he is going inland, he happens to pass over his own lines and is shot down. The other two aircraft, seeing what is going on, panic and break hard left. The first one crashes straight into the sea because they are so low and the other one flies straight into a radar aerial on another ship. Four aircraft for one Sea Cat missile. Who is going to analyse that?

MacDonald

Dr Friedman?

Dr Friedman

Wow! How can one react to that?

I am sure that the Vulcan made the Argentines think. But it's not clear to me that they were deterred from doing anything after all that thinking. Also, I would suggest innocently that if you had your druthers, you would have had CVA-01 and not the Vulcans.

There is no question but that the Argentines were sobered to some extent. It just didn't show up too much in what they didn't do. They did not reserve their air force for home defence. Apparently they flew everything they had. They did not hold back from trying to bomb the ships.

Now let's talk about the Backfires in the North Atlantic. The British have these tiny little escort ships which, of course, are not worth shooting at and the Russians have these submarines. A submarine carries something like 24 torpedoes and a Backfire carries one missile and there aren't that many Backfires. But, of course, those escort ships with the towed arrays are capable of sinking submarines. If I were the Russians, making my force calculations, I'd roll back the escorts and then take out the merchant ships with my subs. The Soviets don't have enough Backfires to take out all those merchant ships. But they have a whole lot of subs that are prevented from breaking out by that patrol line across the North Atlantic. So I would roll it back.

In World War II, you didn't sink escorts when you were attacking convoys largely because there were so many of them. It didn't pay. But right now, the number of escorts is pitifully small. They are very valuable and very capable. It really pays you to roll them back.

Also I would remind you that the official line has not been that we know there is an air threat but they will go after the merchant ships. The official line has been, 'There is no air threat in the North Atlantic; there is an ASW threat.' I can understand your reasoning. But the official line is not like your reasoning. Either there is an official line for us stupid outsiders and another one for those who are sophisticated inside - which leads to lots of trouble in Parliament - or there are those inside who can figure out that the official line may not be completely haywire, but no one outside has quite gotten the message. That's very dangerous. So I stand by what I say.

The Sea Darts kept the Argentine aircraft down low and caused them to have a lot of operational accidents and I'd love to know exactly how many. Still, the fact that they knew that you could saturate Sea Dart is probably why *Coventry* was sunk and once that kind of thing gets around, your ships are going to be in for a very rough time. That sort of thing is now known, but it was carefully concealed for ten years, while the British were trying to sell Sea Darts and Type 42s to everybody they could. How many other things are hidden in the back room, waiting to come to light in the future? How many of the missile tests are completely phony? Sea Slug was advertised as having a 95 per cent kill rate in the British technical press. That's obviously ridiculous, but that is the sort of thing that is published. How many other things are covered by the Official Secrets Act that will cause disasters in some future war? In the United States, our dirty linen tends to get washed in public. A lot of us don't like it, but I think it protects us against some of the more grievous disasters.

Colonel Liston

Dr Friedman has made some very cogent comments on the British ships in the Falklands, referring to their lack of certain capabilities. However, I have read some of the stuff coming out of the United Kingdom where they seem to be concerned about what ministers call 'the road to absurdity,' where costs rise so fast that the numbers you can buy with any reasonable projection of the defence budget decrease to the point where you do not have the numbers of units that are required.

I would ask you to comment on this argument of quantity versus capability or quality. I understand it also rages at times in the United States.

Friedman

It rages everywhere.

In the British case, the real question is whether they face the rising costs and make a public statement to the effect that they need the capability, so they are going to have to pay for it, or whether they try to kid themselves. Type 42 is the cheapest platform that would get Sea Dart to sea. It might have made sense operationally to say 'To hell with area defence altogether. We will live with a much better point defence system.' But what you had with Type 42 was a political attempt to get a weapon to sea so that it could be sold, and so that they could explain why a larger and more expensive destroyer, Type 82, was not being built in numbers. It is obviously true that their budgets are limited. It is obviously true that you have to cut down on certain capabilities. But it is also true that ditching the whole Trident programme might have been a much better way of cutting capability. Maybe they should have bought cruise missiles or something like that. You can't have everything. Since World War II, the British have tried to maintain every possible capability to be a great power. It is very hard for them to face that their economy can't take it. The whole Falklands crisis was about whether or not Britain is still a great power. Others will disagree with me violently. The whole Trident argument is about that. They have to make decisions at some point as to whether they do or do not need certain capabilities. How important is that independent nuclear deterrent? That's a great power issue.

Perhaps they believe that war is so unlikely that they want it much more than they want realistic conventional capability in the North Atlantic. That is a definite decision. I'm worried that the level of defence debate is so low that no one even recognizes that there is a decision to make. Furthermore, within the British navy, the argument between the submarine community and the surface community has been much more violent than the ability of any minister to understand. I think that's why you had those bizarre cuts in the last White Paper. It has to do with the fact that there is not a well informed naval press.

It is not the press's fault. It is the fault of the British government for their attitude that these decisions have to be made inside. A lot of their discussions about what the technology is are patently crazy and have been for many years. They are made by people with no technical ability. I'm thinking of the Sands White Paper in 1957. They deserve what they're getting, but I don't see why people should be dying as a result.

I'll leave it at that. It is political absurdity. We are watching the chickens coming home from twenty or thirty years of political incompetence.

Mr Jean Fournier

My question is directed to Major Elliot. Because I am an ex-foreign service officer, it is a non-military question.

The outcome of the Falklands war has produced a temporary solution to the Falklands political problem. What is the thinking in Whitehall, especially in the FCO, at the present time about future United Kingdom-Argentine negotiations to solve the problem?

Elliot

I'm not sure that I can read the FCO mind. The general feeling around Whitehall is that the FCO has gone back under its rock and is hoping the situation will go away.

Perhaps that is a bit unfair. I think there is no doubt that the United Kingdom is going to be much more difficult to deal with over the Falklands than they were prior to February 27, 1982. The basic principle, that the islanders must have the right to self-determination, will still apply. And they are looking over their shoulders at Gibraltar, which has a parallel problem.

The decision of the United States to not get in the way of the recent Latin American-inspired resolution made the British very unhappy, but there wasn't much they could do about it. It's not going to make them any easier to deal with. I think that the principle is going to be that the islanders must be given a chance to make their case.

I suggest that the islanders themselves have played the thing rather badly. There should have been a tremendous referendum, with all sorts of people invited to see it, to show just how much they are behind whatever line of reasoning the British come up with. The solution might be something parallel to St Pierre and Miquelon or Réunion vis-à-vis the French. Whether that would survive in the United Nations, I don't know.

I'd like, if I may, to go back to a comment this morning on trench foot. I think it is important to remember that there were three different ground force groups in the Falklands: the paras, the marines and the regular infantry. The marines do a lot of Arctic exercises. They are dressed and equipped to live in the cold wet environment of foreshores. The paras have a somewhat different mission - a long range fairly dry environment. Their footwear was not the same as the marines and they suffered more than the marines. Of course the poor unfortunate PBI who rolled in afterwards were not equipped for either kind of environment. They suffered rather more severely from weather conditions. Marching in the Falklands boggy ground means that feet are wet no matter what you've got.

Theodore Olson, York University

Everyone has been grateful to have the Falklands issue fleshed out today, and there have been some surprises. But a question came up at lunch which I think is of some interest. How much weight of analysis should be put on this single event with regard to high tech materiel? We would be much better off if we had a comparative situation with much of the same equipment in use. There is a large war, that has been fought over the last year and a half, that has received extremely little analysis in the documents available to most of us. That, of course, is the Iran-Iraq war. But has the panel any suggestions as to why such a minute amount of attention is being paid to that war?

Charters

We have been following this conflict for some time, and it is my understanding that both countries have made it a policy to keep newsmen out of the battle area. So unless a newsman is willing to risk his life to cover it covertly, there is not much coming out in the form of journalism. It is hard to build a data base on which to do analysis.

Elliot

I'm glad to hear that the other organization in this field has the same problem that mine does, and for exactly the same reasons. There is just very little coverage. Even when we do get somebody back from there, he usually does not have the expertise or knowledge to interpret what he has seen. Anybody near any military unit is pretty well circumscribed. Even people coming out of the Iranian element are very limited. You do hear suggestions that perhaps the revolutionary guard have split the responsibility with the army and since the revolutionary guard don't know much about soldiering they are doing all sorts of stupid things.

But there is no ability to analyse the mechanics of war. So far as we were concerned, the general feeling at the beginning was that the army middle management had been pretty well abolished as a result of the coup and were not capable of putting a reasoned offensive together. How do you explain the various offences that have taken place? You have two choices. Either you say they have got their act together or you say that the Iraqis have folded and gone home - which is entirely possible. But, as Dr Charters said, there is nothing of really significant evidential basis to hang your hat on.

Friedman

Don't forget something else. For some reason, wars between Third World countries tend in general to be of much less interest to western journalists and are not very thoroughly covered. The example I would give is of a gigantic war in the Sudan for about seventeen years that had almost no coverage. You are dealing with two countries that are well able to prevent anyone from finding anything out; you are dealing with a war that is large and difficult to make sense of; both countries are interested in reporting enormous successes and preventing you from finding out the facts. They don't get anything out of having journalists and they don't have any internal pressure to have the war reported. In any war involving a Western country, there is enormous pressure from those back home to find out what is going on. Most Canadians, Americans, and British are very interested to know what is happening when Britain goes to war. So there will be a lot of coverage and there will be pressure on the British government to give coverage, even if they don't want to.

Elliot

One final comment. It does raise an interesting question about the extent to which our own perceptions and even our own priorities for strategic analysis are set by the happenstances of journalism!

Tim Descheneau

The high technology field of computing has promised since 1950 revolutionary new ways to reveal the truth to mortals. This usually appears on delivery as nice TV pictures and neatly printed reports. What they are doing is mechanizing very simple activities so that you can do a lot of them in a short space of time. The world's two most labour-intensive activities are planting rice and programming computers. I predict that if someone thought up a wholly new way of thinking about things and wanted to put it on a computer, it would take at least ten years to programme it and get it right.

My second comment is what I call the 'great blue pumpkin syndrome.' IBM found out a long time ago that if they charged two million dollars for a box, the customer will be incredibly disappointed if the salesman brings it in and pulls it out of his pocket. Therefore their computers come in large arrays of blue boxes, the contents of which turn out to be mostly air. A comment was made this morning that artillery fire-control computers were left behind because they were too hard to move. I suggest that the fire-control computer for an ordinary artillery battery should fit in your shirt pocket. The display might be a bit larger, and they should stop driving that model with steam!

GORDON MACNABB

CREATING THE MINDPOWER BASE OF HIGH TECH

Mr Chairman, ladies and gentlemen: The last time I spoke to a meeting organized by the Canadian Institute of Strategic Studies I addressed, in my capacity as deputy minister of energy, the challenges facing that critical element of our national economy. At the outset I want to say to you that I view the subject matter of my comments this evening – Canada's manpower and mindpower – to be every bit as critical to the future of this nation and of even greater significance in the broad definition of the word 'strategy.'

Before I embark upon a defence of this position, and suggest where the challenges and opportunities lie, let me very briefly describe the perspective from which my views are drawn. I currently head the federal crown corporation called 'NSERC,' which is the largest single source of support for university-based research and for scholarships and fellowships for graduate study. You will not be surprised, therefore, to learn that my comments will focus on our graduate schools, and even more particularly on certain disciplines within those schools. NSERC, the acronym for a long bureaucratic name for my council, currently has an annual budget of $226 million. Last year, we provided research support of some nature to over 17 000 individuals, and included in that total were over 3000 scholarships or similar scholastic awards. It should be obvious that through these programmes, and through the input of hundreds of volunteers on our committees, we can monitor quite closely the health and vitality of research training in our universities.

With this as background, let me turn to the issue at hand. It is certainly not a new one, as manpower training has been an item of concern for a long time. It is hard, for example, to improve upon the comment of Benjamin Disraeli in the British parliament in 1874 when he said, 'Upon the education of the people of this country, the fate of this country depends.' Seventy-five years later H. G. Wells observed that 'human history becomes more and more a race between education and catastrophy.' That was thirty years ago. His observation surely is even more pertinent today.

You have been discussing the high technology war of the Falkland Islands today and therefore I am sure that you understand fully the race of technology with regard to military activity. However, that race is also a critical element of almost all economic developments, from the more basic ones of resource development and farming to the exploitation of new developments in microelectronics and biotechnology.

Are we equipped to run in the race, let alone with it? I will try to condense my comments by breaking them down into the good news and the bad news. So as to end on an up-beat, let me give you the bad part of the report card first.

The graduate schools of Canada represent the apex of our pyramid of learning, and strength or weakness at that point dictates the overall viability of the structure itself. In more military terms, those schools represent the spearhead of a broad frontal assault, where success is a critical factor in the overall operation.

You would think, therefore, that those graduate schools would be the focus of attention and of attraction in our educational programmes. They are not. In fact, graduate training in some very key disciplines has only survived during the past decade as a result of an influx of visa students.

Consider for a minute the record of some of the disciplines so critical to us in this increasingly high-technology age:

Ten years ago there were 611 Canadian and permanent resident students enrolled in full-time graduate training in electrical engineering. By 1979 that number had dropped to 376. In other words we are looking at less than a hundred master's and doctoral graduates per year in this critical discipline by the mid-1980s. That is hardly the way to win a technological race.

Chemical engineering was no better; 322 students were enrolled in 1972 and 188 in 1979 – for a potential supply of less than fifty new graduates per year.

Everyone acknowledges that our ability as a nation to grow and to develop and compete internationally is becoming increasingly dependent upon our systems engineering and computer science competence, but you can't tell that by the number of Canadians taking advanced research training in our graduate schools in those disciplines. There were 448 students enrolled in 1972 and 332 in 1979. That's not an advance; it's a retreat. Like so many other things, non-Canadians seem to appreciate our opportunities more than we do. In the eight-year span I have been speaking of, visa student graduate enrolment in computer science went from 83 to 173, or more than doubled.

These visa students must leave Canada when they graduate. While we can aspire to attract some of them back, that comfortable habit of the past will not be as productive in the future. These graduates are in great demand everywhere in the world and while they may leave with a warm attachment to our country, that by itself is no assurance of their return. In any event, our heavy dependence upon a reverse brain drain for such a vital resource is not acceptable from a self-sufficiency point of view, nor from our concern for the economy of the home countries of many of those students.

Quickly let me skim some of the other aspects of the bad news.

In the universities themselves, eroding budgets and the simple lack of too few highly trained people has greatly hindered staffing in the engineering and computer science faculties. Universities have not been able to compete with industry for the talent that has been available and therefore their ability to train at both the graduate and undergraduate levels has suffered. This 'eating of our seed corn' by industry has also happened in the United States, but there at least there has been a recent move to greater university/industry co-operation and joint funding to help correct the situation.

Our academic community itself is characterized by a great concentration of staff in a fifteen-year age grouping and that group will start to reach retirement age in eight to ten years. In the 1990s we will need very large numbers of well trained young faculty. The beginnings of that supply should be entering our training system now.

Next to universities, government research departments are the largest users of Master's and Ph.Ds and they too suffer from the problems of an aging research force. While the situation within the civil portion of the Department of National Defence is not as serious as that at Agriculture Canada, recent studies would suggest that about thirty per cent of DND's engineering and scientific staff is over fifty years of age.

When dealing with all of these numbers and statistics, it is easy to get caught up in a mathematical exercise and therefore to limit one's focus to the challenge of replacement eight or ten years down the road. What happens in the interval, however, is of equal importance. How productive, how current, how stimulating and stimulated is the aging core of researchers going to be in both the universities and government departments when they lack the catalyst of a steady influx of bright young recruits? There are exceptions, of course, to all rules: but, in general, scientific output drops in the later years of a scientist's career. As a result, unless there is an easing of staffing freezes, we face a strong possibility of eroding scientific output in both university and government laboratories over the next two decades. Heaven help us if a large majority of the current research staff insist upon a right to occupy their positions after the normal retirement age of 65.

That's enough of the dismal facts. Let me now turn to the good news. Here I start at the undergraduate level - the recruits of the future, if you like. The picture over the 1970s was radically different from that of our graduate schools, especially in the three critical disciplines I have talked about. Consider these facts:

Enrolment in electrical engineering of Canadians and permanent residents grew from 2940 in 1972 to 4939 in 1979, a growth of 65 per cent. By 1981 it had grown by another 14 per cent to over 5600 students.

Undergraduate enrolment in chemical engineering grew from 1276 to 2900 during the same period - a growth of 128 per cent. Computer science enrolment grew from 2518 in 1972 to 3860 in 1979 and to over 6000 students last year, a growth of 244 per cent in ten years.

Clearly we now have the numbers required for a rejuvenation of our graduate schools and I can vouch for the quality of the candidates who will be presenting themselves. You have to be the equivalent of an Ontario Scholar these days to get admitted to these undergraduate courses and one of the joys of my job is to see the quality of the survivors at the bachelor's level.

A second area of promise is that we have managed to stop the decline of graduate school enrolment in the last three years and have seen increases of twenty to thirty per cent between 1979 and 1981. A new national commitment to research and development, and increased federal support through vehicles such as my council were contributing factors to this modest recovery.

What is even more encouraging, however, is that in this *current* academic year it seems that there has been a dramatic upturn in interest in graduate training. All the facts are not in, but at the University of Waterloo, for example, graduate enrolment in computer science and electrical engineering (by Canadians and permanent residents) is up by thirty per cent. At the Technical University of Nova Scotia graduate enrolment is up by about forty-five per cent.

NF?

There is no doubt that a major contributing force to this renewed interest in graduate training is the current state of the economy and the loss of immediate job prospects. But no matter what the cause, we are being presented with a unique opportunity to tap the rapidly increasing numbers of graduates at the bachelor's level and to build our research competence through graduate training within the universities. It is a chance to invest in our long-term technological future. Such an opportunity to regroup and to reorganize may never come our way again and all logic says that we should seize it while we can. But will we? Will the same fiscal situation which created the opportunity also prevent the requisite action to implement the recovery? Personally, I am confident it won't – at least not at the federal level. My council has received significant real increases in funding over the past three years and the growing appreciation of the necessity of R&D for long-term economic health augurs well for the continued growth of such support.

However, the required increases in scholarships, scientific equipment and research funding goes well beyond any perceived level of support NSERC can provide. Joint effort and funding will be needed: provincial input is necessary, as is that from industry. In particular I stress the latter. Industry will need an increasing number of the products of our graduate schools in the years to come and it behooves them to join with academic staff in coping with the increasing teaching burden, in helping my council through the joint funding of high-cost research equipment, and by providing direct scholarship support for those disciplines that are their highest priorities. Funding alone, of course, would help; but the visibility of industry support to the graduate-student population would be equally important as a clear sign of recognition by industry of the value of postgraduate research training.

In summary, therefore, ladies and gentlemen, we have not been preparing ourselves well for the technological opportunities and battles of the 1980s and 1990s. However, a welcome by-product of an unwelcome period of recession, the silver lining of the dark economic clouds, may well give us an opportunity to recover some lost ground. We have to accept the fact that it takes time to build a stable base for recovery and long-term economic growth since the increased supply of researchers at the doctoral level will not enter the work force until the latter part of this decade. But better late than never.

To produce the technologies and the economic growth that we all want will entail a total commitment to the health and quality of research training in our universities. That can happen and I believe it will happen.

TERRENCE LISTON

THE IMPACT OF HIGH TECH ON MILITARY MANNING

LCol Brian MacDonald

Today we move on to a further examination of the human dimension of this general issue of 'War in the 80s: Men Against High Tech.' At this point we look at the problems of the manpower planners in dealing with manning the forces in an environment in which high technology has become a dominant characteristic.

We are very fortunate this morning to have Colonel Terrence Liston with us as the first speaker. He was born and raised in Montreal and then joined the Royal 22nd Regiment of Foot in 1957. He served in Quebec, in the Congo and in Germany, and in Germany commanded the first battalion of that regiment. He has spent three years as the deputy Chief of Staff for combat development at Mobile Command headquarters, and in his present position is the senior policy analyst under the chief of policy planning in the office of the ADM (policy) at National Defence headquarters.

Colonel Liston is a graduate of Queen's University at Kingston. He has studied at L'École Nationale d'Administration Publique in Quebec, is a graduate of the Canadian Army Staff College and the National Defence College. Colonel Liston.

Colonel Liston

It was with some anxiety that I accepted the invitation of our chairman to inject a discussion of personnel questions into a seminar which focussed on technology. It was heightened when I heard that General Burns, who has written the standard text on the subject, would be in the audience. And when the issue of 'national service' showed up prominently as one of the alternatives in the advertised description of my talk, I must admit that my anxiety came close to panic.

Nevertheless, an exercise which involves looking into the future and examining alternative courses of action to deal with both technology and demography must go beyond approved policy and does not necessarily reflect the current views of the Department of National Defence. I am grateful that the Department has so generously accepted my speaking to you this morning, outside the framework of approved policy, so as to provide a basis for some of your discussion.

The description of this talk – 'volunteer forces or national service' – represents alternative solutions to a problem. Our discussion of these solutions will perhaps be more relevant if we are able to agree on the nature of that problem. To this end, I will spend the first part of my talk in establishing a number of assumptions in regard to the impact of high technology on our forces in the future, the need for and the availability of manpower, and the requirement for skill and knowledge. With this definition of the problem well established, we will then be better able to examine the different means of providing the necessary quantity and quality of personnel.

THE PROBLEM

Technology

Technology, which I will examine first, impacts on our posture within the North Atlantic Alliance and on the type of battle that we must prepare for.

NATO has, since its formation, sought to find inexpensive solutions to the security of the Western world. In the early days, nuclear superiority seemed to promise such a solution, compensating with minimal effort for superior Soviet conventional strength. In the last two decades, however, the certainty that strategic and theatre nuclear weapons would deter conventional aggression by the Warsaw Pact has been the subject of increasing doubt; in an environment where there is nuclear parity the threat of the use of nuclear weapons by one side or the other is much less convincing.[1] On the other hand, although we have for the past 15 years paid lip service to the conventional component of our official strategy of 'flexible response,' General Bernard Rogers, the Supreme Allied Commander Europe, has argued that the inadequacy of alliance conventional capabilities leaves NATO in a posture that 'in reality can only support a strategy more accurately labelled "delayed tripwire" ... at best sufficient to allow NATO only the time and security necessary to deliberate and escalate to the use of nuclear weapons' in the face of a large-scale conventional aggression.[2] In the words of Senator Sam Nunn, 'There is a slowly emerging consensus running across the political spectrum - from defence-oriented conservatives to the more responsible segments of the anti-nuclear movement - that it would be desirable to reduce this reliance on nuclear weapons and to strengthen NATO's conventional forces as a means of deterring non-nuclear Soviet aggression in Europe.'[3]

Technology may allow this to happen. The lethality of conventional weapons is being increased to the point that their potential is approaching the effect of small battlefield nuclear weapons; this, combined with the advent of 'high-tech' micro-electronics and sensors, has greatly improved the ability of units to accurately acquire targets and deliver large amounts of conventional fire power onto them. This enhancement of fire power vis-à-vis manoeuvre will tend to favour the defence and suggests that a strategy of increased dependence on conventional forward defence in Europe is possible without necessarily matching the Warsaw Pact forces, tank for tank. With this in mind, General Rogers has asked for a capability which would allow him to hold the lead divisions of a major Warsaw Pact conventional attack, while other conventional forces based on this new technology are used to destroy in depth the enemy's follow-on echelons, to counterattack his forward elements and to disrupt his radar and communications. Simultaneously, the European countries are being urged to make more use of their vast reserves to provide low tech defence in depth.

In this strategy, our conventional forces in Europe, and those earmarked for Europe, must be able to survive longer than only the time needed to trigger a rapid nuclear escalation; their land forces would contribute to the holding and counterattacking of the lead Warsaw Pact divisions, while our air force contributes to stopping the follow-up echelons and our navy helps keep the sea lanes open. Most significantly if deterrence is to be effective, we must be seen to be prepared to sustain our forces in this intense conventional combat for whatever time would be necessary to reverse a conventional attack on the Alliance.

Personnel Requirement

Within this technological scenario we can now ask one of the questions I was requested to address. Will the aging of the Canadian population and the shrinking of the available pool of military age affect our ability to raise the required numbers ... as it will in Germany, the United States and many other countries? Let us identify the numbers of personnel, and the degree of readiness, that could be required in order to fulfill the needs of the various missions of the Canadian

Forces. This will give us a rough idea of the requirement for standing forces as well as for the various types of reserves.

Certain missions require, in peacetime, standing forces, prepared to move rapidly to their likely battle positions. Their members may be long-service professionals or shorter service personnel who have been trained and integrated into their unit. In Canada's case, these missions include mainly our units stationed in Europe, the mobile land and air forces earmarked to rapidly reinforce the NATO defences on the northern flank of Europe, our NORAD surveillance and interceptor forces and those of our maritime forces which carry out surveillance in peace and have an immediate reaction role on the high seas in war.

If, however, we cannot afford such levels in our standing forces, the next best thing is to have reserves of fully trained personnel who can fill in the vacant positions in standing units and who are available for recall from civilian life. Canada's Supplementary Reserve, as Mr Harquail's SCEAND report shows, has been up to now a list of names of people, many of them unfit, who have had some previous service with the regulars, the reserves or the cadet services. The Canadian Forces are now in the process of expanding this list, with some degree of success. The Primary Reserve is also tasked, in Canada, to provide such individual augmentation and in many trades this can be usefully done. However, in the army in particular, the main basis for bringing the standing forces to full war strength should be a ready reserve of fully trained, fit and available individual augmentees similar to the demobilized conscripts of continental Europe or the ex-regulars of the United Kingdom Regular Reserve and the United States Individual Ready Reserve.

Missions which will continue to permit a somewhat less critical level of readiness in the foreseeable future are those related to the land and maritime defence of Canadian or North American territory. It is for territorial defence of this nature that national militias such as those of Sweden and Switzerland exist. Other countries develop units of volunteers such as the United Kingdom Volunteer Reserves, which includes the Territorial Army, or the United States Selected Reserves, made up on the National Guard, the Army Reserve, the Air Reserve, etc. In Canada, the maritime tasks of protecting Canadian ports and the army mission of defending North American territory could be met in large part by units of the Primary Reserves.

Finally, there are missions which consist of rather simple, although important, support functions such as defending the local community and guarding vital points. Many continental countries have a form of home guard which relieves the regular and reserve forces of this function. In 1982, even the United Kingdom decided to form a new home guard of about 4500 volunteers.[4] Canada has a large number of vital points and military installations which, in a crisis, would require at least 10 000 guards and we should, I believe, identify this mission as being in a special category, requiring levels of training and equipment appropriate to the somewhat limited task.

Considering all these missions together we can conclude that, in the first few days of an emergency, the Canadian Forces would require over 100 000 troops. The 1982 Senate report on manpower identifies a war establishment of 108 000 and recommends that 91 800 be in the regular force. The greatest weakness is in the land forces, where the brigade in Germany is lacking up to 4400 troops from a level felt to be viable, where the land commitments to Norway and Denmark are now met with the same forces, and where only 2000 to 3000 trained combat troops would remain as the nucleus for defending coastal and northern Canada, protecting vital installations across the country and providing aid to the civil power.[5] The Reserves are, as we know, unable to make up this shortfall.

I have not yet addressed, however, the most significant factor flowing from technological change, the need to sustain these military commitments during weeks, and perhaps months, of intense conventional combat. Although the United States is reported to have committed itself to building up from a current level of 30 days sustainability to 'a minimum of 60 days ... in all theaters and at sea by the end of fiscal year 1987 ... with an objective of attaining a 180-day level in ten years,'[6] Canada's position is less precise. The Senate manpower study notes that, should a war in Europe continue beyond 'the initial stage of a week or so ... Mobile Command must be able to draw on some surplus of trained manpower for reinforcement purposes.'[7] The Minister of National Defence, the Honourable J. Gilles Lamontagne, indicated in a speech to this Institute on 28 May 1982 that 'the strengths required to enable Canada's units and formations to be viable at the outbreak of hostilities, and in subsequent battle, have only been identified in the past year ... The Department of National Defence, in a separate submission to Cabinet ... intends to outline what it needs, in terms of personnel, operations and maintenance, and capital procurement, in order to viably meet and sustain Canada's commitment.'[8] Let us, in the meantime, make our own estimate as to what the need could be.

In discussing the replacement of casualties we must be clear as to what sectors of the forces will be affected. Assuming that our ships and aircraft are manned at the outbreak of hostilities, one is more likely to be left with a surplus of crews as airplanes and ships are lost and damaged. On the other hand, much land force materiel survives the shrapnel and armour-piercing shells which are so destructive to personnel. Much of it is inexpensive enough to allow it to be stored in reserve to re-equip depleted units. It is therefore in the land forces that the replacement of casualties will be most critical to Canada.

It would also be useful to look for indications as to the numbers of casualties that we should replace. The most recent conflict of a high-intensity, mechanized nature similar to what could be expected in Europe was the 19-day Yom Kippur War. Barker describes the actions of Colonel Reshef's 14th Armoured Brigade in preparing for the recrossing of the Suez Canal: 'By the time the battle was over his brigade had suffered nearly 100 per cent casualties – for the third time since Yom Kippur.'[9] Tanks can be, and were, repaired in a matter of hours; but men have to be replaced. Moreover in an intensive battle the number of psychiatric casualties ... what was known as shell shock in World War I ... is a further cause of losses; in hard-hit units they could amount to over 50 per cent of unit strength in a short time.[10]

Canada's forces deployed in Europe would be thrown immediately, from a peacetime situation into 24-hours-a-day high-intensity battle worse than that experienced in the Middle East. The need to replace or relieve individuals, units and whole brigades could require a total force of the order of 150 000 to 200 000 trained personnel in order to meet and sustain Canada's current commitments for the first few months of combat. It would not be unusual to expect that, as in European armies, these forces would include a large proportion of reserve units. Even in the United States, the Selected Reserves are tasked to provide 'nearly half of the strategic military airlift and land combat power, and two-thirds of its logistics support capability. The Active Army relies on the Reserve Components to provide the preponderance of combat support and combat service support units for major operation plans.'[11] Finally, if we were in the long term to pursue a six-month capability, we would have to think in terms of a major expansion, which requires the recruiting and training of civilians. Let us assume, solely for discussion purposes, that it would push the total force requirement over the 500 000 level.

It can therefore be suggested that, for purposes of today's discussion, and assuming modestly expanding defence funding, Canada's total personnel requirements should include:

1 standing forces of close to 100 000;
2 a ready reserve of individual ex-service men to fill gaps in establishments and to provide trained replacements in the first weeks of fighting; the replacements would be, in large part, junior combat arms personnel and could number up to 25 000;
3 reserves of trained units which would be immediately mobilized, mainly to defend North America and to sustain the Canadian land forces overseas; this could represent a further 40 000 to 60 000;
4 a home guard to provide for the protection of vital points: at least 10 000; and

5 up to 300 000 civilians who would be recruited into an existing infrastructure of mobilized reserve units so as to permit the long term sustainment of the war effort.

Will the aging of the Canadian population in the 1980s and 1990s make these numbers unattainable? In 1981 the 17- to 24-years-of-age 'cohort' of young male and female Canadians was 3.7 million; by 1991 the decline will be 21 per cent to a level of 2.9 million. If we consider only the males in this group, their numbers will decline from 1.9 million in 1981 to 1.5 million in 1991.[12] In the 1970s Canada has been recruiting approximately one per cent of this male cohort annually, into the Regular Force. To increase the intake from 20 000 to 30 000 recruits annually in 1991 it will be necessary to recruit about two per cent of the male cohort. Thus the maintenance of a standing force of 100 000 will be more difficult but not impossible.

Let us assume, however, that it will be necessary to expand the forces to 500 000 in a crisis in 1991 and that the whole intake would come from 17- to 24-year-old males; about 26 per cent of the cohort would be required. This would still represent a considerably smaller portion of the eligible manpower being trained even in peacetime by many of our NATO allies. For example, Norway inducts 97 per cent of its 19-year-old males each year and Germany calls up 50 per cent. When we consider that many positions will be filled by females and older personnel it must be concluded that the required numbers to meet Canada's defence commitments will continue to be available into the 1990s.

Skills and Knowledge

Accepting that we will be able to find the necessary numbers within Canadian society, many wonder whether they will be the right kind of people to cope with high tech systems. Let us examine the impact of new technology.

Operators at the lowest levels must, without necessarily being technologists, be more intelligent, responsible and highly trained than ever before in order to cope with the unprecedented lethality and range of these new systems. However, much of the work involved in acquiring targets, estimating ranges and guiding weapons to the target will be done automatically and with great accuracy. When the first anti-tank missiles appeared in the 1960s, the Canadian Army demanded very tight selection for their operators. Now with the current generation of guided weapons, virtually any soldier can use them. In a similar vein, the F-18 is relatively simple to fly and use in combat. Where the complexity is excessive it flows in large part from a failure to consider the 'people requirements' early in the design process. Engineers must design equipment not only to meet the limitations of our Regular Force recruits, but more particularly, the limited training time and facilities which will be available to Reserves.

The impact of high tech on the maintainer will be simplified by built-in test equipment (BITE), automatic test equipment (ATE), automatic switchover to available spares in case of failure, repair through replacement of throw-away printed circuit boards, and so on. Maintenance in the forward area could involve simply replacing components rather than repair and this could well be the way of the future. In practice, in the Canadian Forces, the trend is towards a re-structure of trades so that the first-line operators and maintainers benefit from the increased speed and simplicity of the new technology, while another level of highly-skilled technicians is being trained to, for example, repair the 'throw away' printed circuit boards. In the case of the F-18, the forces will actually be doing maintenance which, on previous aircraft, was done by factory contract. In short, the advent of high tech has indeed resulted in a need for some highly skilled technicians capable of diagnosing 'systems' problems, reprogramming and so on, but it has also made first line maintenance fast and relatively simple.

There is in any case no difficulty in finding personnel with the aptitude to carry out these technical functions. We have had, in fact, an educational revolution in Canada. In 1957 more than half of Canada's youth still left school at grade eight, nine or ten; in the 1980s more than 60 per cent of youth will finish high school. Moreover post-secondary education in the technical and community colleges has flourished across the country and this trend will continue into the 1980s. Those with university education should double from 6.9 per cent of the population in 1966 to 12.5 per cent in 1986. It is even expected that, in Canada and the United States, there will be a surplus of university graduates.[13]

Our problem, then, is simply that our recruiting pool is becoming extinct in the wake of our educational revolution. High school and post-secondary graduates will soon make up 60 per cent of the population while our recruiting, our training system, and our organizational ethos is aimed at a World War II society, 75 per cent of which had not finished high school. Our main personnel problem in the age of high tech will not be one of finding electronics technicians; it will be that of recruiting into the operational elements the thousands of regulars and reservists which we need from an educated society which has the highest expectations of career, status and money.

THE SOLUTIONS

Let me, before focussing on possible solutions, summarize our definition of the problem that flows from the factors of technology, demography and education:

1 first, the impact of high technology on Alliance strategy could well lead to a need to sustain our commitments in conventional combat for an extended period;

2 secondly, an appropriate force model could call for a small increase in standing forces, and a major improvement in the size and training of the reserves, largely in the land forces ... and well within the capability of Canada's decreasing military age population;

3 thirdly, the skill and knowledge required of both operators and maintainers is within the capability of our youth; and

4 finally, the educational revolution which is taking place is eliminating the traditional recruiting base of volunteers for the forces.

In short, we must find ways of recruiting those with high school education, and better, into the operational sectors of both the Regular Force and the Reserves.

There are four factors which, either alone or in combination, could contribute to a workable solution: compulsion; pay and benefits; organizational climate, and what I have called legitimacy.

Compulsion

For many nations, conscription and a citizen army are essential characteristics of a free society and the tradition may date back several centuries. Conscription, in most systems, ensures that all classes of society bear the price of defence. The armed forces are, in their eyes, the 'school of the nation' and compulsory military service is a necessary, if at times inconvenient, part of reaching adulthood. This is not, however, the Canadian tradition. MGen the Earl of Dundonald, the General Officer Commanding Canada's Militia at the turn of the century, considered compulsory service to be 'un-British.' Subsequently, the perceived need to briefly resort to conscription for overseas service when the voluntary system faltered in two world wars led to serious cleavages within the country which have perhaps not yet been completely healed. Although various forms of peacetime conscription have been regularly advocated by the Conference of Defence Associations, they have made no headway at all and the issue has not been pursued since 1957. The diminished threat, the lack of public support, the cost 'and the fact that, in the main, the voluntary system had worked' are the reasons invoked for this Canadian reluctance.[14] Although Canadian law, until 1950, paid lip service to a 'levée-en-masse,' Professor George Stanley has suggested that from 1855, the volunteer system has been the fundamental feature of Canadian defence policy.[15] In the light of this tradition, and of our relatively modest manpower requirements, there is little point in giving much consideration to either the levée-en-mass or to compulsory national service in standing forces of the French or German type in peacetime.

A short compulsory training period in order to create trained reserves, where we have the greatest need, would probably be the type of peacetime conscription which would be the most acceptable. It was such a measure which was introduced in Canada in October 1940 with relatively little difficulty and which was advocated regularly by the Conference of Defence Associations in the 1940s and 1950s. It has also recently been proposed by the United States Comptroller General as one of the options to improving the United States manpower situation, although the United States Department of Defence rejected it as being too costly and providing inadequate training for men who would be thrust into battle within 30 days of mobilization.[16] It is nevertheless a proposal that we should not reject out of hand.

Finally, there is the possibility of instituting some form of compulsory service only on mobilization, as is currently proposed in the United States. This measure would allow the quick call-up, in an orderly and phased way, of the several hundred thousand men and women needed to sustain our commitments beyond a period of a few months. It would, however, require the registration of young people in peacetime, which is an obvious obstacle.

Let us leave these compulsory schemes and turn towards the voluntary system of recruiting and retention upon which our peacetime forces have been based in the past.

Pay and Benefits

In voluntary systems pay and benefits would appear to be one of the principal determinants of recruiting and retention success in peacetime. The presence of well qualified high-tech specialists in the forces will make them particularly vulnerable to raids from industry if the pay and benefits we offer them are not comparable. At the same time operational jobs in the forces which are physically and psychologically demanding and which often offer few concomitant benefits or marketable skills will exert little relative attraction to the high school graduate interested in a trade, status and a reasonably comfortable life. These phenomena have become strikingly obvious in the United States where with the reversion to an all-volunteer force, and (in the late 1970s) a poor pay structure, the armed services were unable to attract either the required quantity or the needed quality of personnel nor were they able to retain the technically trained officers and NCOs they needed. However, a 25 per cent raise over two years in 1981 and bonuses of as much as $8000 for entering a combat arm, combined with a faltering economy, have allowed the United States forces to meet all their quotas and to increase their intake of high school graduate recruits from 55 per cent in 1980 to 80 per cent in 1981.[17]

But pay alone will be insufficient to attract and retain, in the operational sectors of the forces, the well educated and demanding youths of the 1980s and 1990s, particularly when the economy gets going again. These sectors will have to offer a package of benefits that appeals to the high school graduate's expectations of having a 'trade' and significant security. In the United States, the GI Bill has been reinstituted, giving educational benefits to members who have served six months or more. Similarly in Denmark, regular soldiers earn twelve weeks of civil education per year.[18]

In Canada, the desire of combat arms recruits for a marketable trade has been recognized through the institution of the Land Operations Trade Reassignment Program (LOTRP) which allows combat arms soldiers to transfer to one of 47 non-operational trades any time after three years of service. This has been a useful incentive. Some suggest that all non-operational trades should be manned from the operational sectors of the forces so as to make the best use of this 'benefit' in recruiting. It, as well, provides very well motivated specialists who know their raison d'être.

On the other hand we have seen that one of our important goals is to produce a large supplementary reserve, and that we will have to find its members in the mainstream of Canada's educated youth. To this end, we should consider, for the operational sector, a short contract of two years or less in the Regular Force, with relatively modest pay but with substantional post-service benefits likely to appeal to a high school graduate, provided he accepts a commitment to be immediately available in the Supplementary Reserves. Benefits such as subsidized post-secondary education, low-cost mortgages and preference in government employment come immediately to mind.

The Primary Reserve, if it is to expand to 40 000 or more, and provide proper training, must also offer a mini-package of benefits to its members. There should be a programme, with pay and attractive benefits, such as tax deductions, which would attract well educated and upwardly mobile youths to proper full-time training (perhaps over two summers, or conversely during one academic year away from school) followed by a commitment to a reserve unit. (In my view, reserve training on weekends is useful at the collective levels but the basic skills must in the future be given on a full-time course if reservists are to be sent immediately into combat on mobilization.) Militia Individual Training and Career Profile (MITCP) should, of course, be continued for those who have been missed by this system, but my proposal is directed at the great majority of the new recruits who have to be brought into the reserves. There should then be incentives to employers, such as tax write-offs, to allow reservists time off to train.

The difficulty with these suggestions is the apparent cost of providing the pay and benefits. However, a great many government programmes provide post-secondary education funding, vocational training and job creation with no *quid pro quo*; for example, since 1971 the Canadian government has spent $3.5 billion on 380 355 person years of job creation, alone. A proposal to tie some of such social programmes to service to the country appears reasonable.

In the United States the notion of 'voluntary' national service has been frequently proposed as a means of offering these benefits not only for military service, but also for civilian service to one's country. This overcomes the opposition of some of those who feel that it is unfair to require military service in return for social benefits. So far it has been rejected by the United States Department of Defence as being far too costly and unjustifiable in terms of defence needs alone. However, in a period of economic difficulty, particularly for the country's youth, it might be a better way of distributing the sums used to alleviate the effects of unemployment. If such a system was perceived as having other benefits for society, there is no doubt that it would contribute to solving Canada's military manpower problem.

Organizational Climate

A more abstract factor governing voluntary recruitment and retention, particularly in peacetime, is the nature of the ethos or organizational climate of the military. Is military service a vocation or simply a job? There appears to be universal agreement that the nature of military society has changed, becoming much more technical, more specialized, more comparable to and integrated with civilian professions. What is disputed is whether this is good or bad; inevitable or reversible.

This discussion is reflected in the categorization of approaches by sociologists into the military 'vocational' approach on the one hand, and the civilianized 'occupational' approach on the other.[19] Some tend to see these categories as distinguishing between a traditional 'mass' conscript army's approach versus that of a highly technical, specialized and voluntary navy and air force.

However in the words of a United States Air Force officer, 'there are perhaps 30 or 40 air forces ... as we move further from the flight line toward support areas, the vocational model recedes and the occupational model appears more prevalent.'[20] Functional specialization, where individuals must perform narrow and specialized tasks in great depth, far from battle, leads to an occupational orientation. A different leadership style, with expertise rather than rank as a primary criterion, tends to evolve in all three services where such conditions apply. The group discipline, firm leadership and mutual loyalty and identity of the fighting elements appears either old-fashioned and irritating, or simply amusing and irrelevant, in this type of environment.

To recruit and retain these skilled technicians, the Canadian Forces will have to seriously re-examine its present system of starting recruits at the 'unskilled' level and moving them in lock-step through predetermined training programmes. We are, in effect, competing at great expense with the post-secondary civilian institutions for trainees; this competition will become more severe as the colleges continue to expand while the youth population diminishes. At the same time, we are unable to accept trained technicians without starting them at the basic level. A frequent proposal is to institute 'lateral entry,' whereby skilled technicians could enter the forces after being trained at someone else's expense. Another possibility consists of using civilian facilities for service members; for example there is a pilot project under way to train marine engineering technicians at St Lawrence College in Cornwall. In-service training must also become more flexible and individualized, using individual computer-managed learning systems with self-pacing material, not only at the training schools but also at the bases where continuation training can be pursued. Indeed the training of F-18 technicians at Cold Lake reflects this approach in part. This is admittedly an 'occupational' approach and is undoubtedly the most suitable for the technical sector.

It is not, however, suitable for the operational sector. The occupational approach does not measure the combat arms or the sea trades by any criterion but marketable skill and knowledge; the tank drivers or anti-tank gunners are viewed as unskilled labour and their status within the trade structure is rated accordingly. Values such as dedication, sacrifice and duty tend to be viewed as naive and old-fashioned. Canada does not, however, have sufficient 'unskilled' labour to meet the needs of the Regular Force and, in particular, the Reserves; nor can any country seriously expect that its least privileged citizens will bear the main sacrifice of war. We must rediscover the ethos that considers the most dangerous and demanding jobs in war to be the most honourable ones, regardless of their technical content. It is in this way, as well, that the operational elements of both the Regular Force and the Reserves will be able to more credibly recruit, in peacetime, from the mainstream of secondary and post-secondary graduates.

The nature of future conflict, with small, widely spread units, suffering high levels of casualties in a battle which maintains its intensity both day and night, will lead to even greater need on the part of the combatants for the qualities of high morale and cohesiveness which are associated with a 'vocational' orientation. The difficulties experienced in Vietnam have led to a growing self-criticism of the organization and values which characterized the United States fighting elements over the past decades and a recognition of the continued need for traditional military qualities even on the modern battlefield. Gabriel and Savage were particularly critical of man-for-man rotation, careerism and the lack of cohesiveness of the United States army in Vietnam. 'By every historical measure of military cohesiveness, the American Army in Vietnam was in a state of advancing disintegration at the time of disengagement.'[21] They blamed, in large part, the officer corps and recommended the re-establishment of traditional values.

The United States has, in the past two years, been making significant progress towards the 'New Manning System' whose objective is to keep soldiers in their units longer to train and serve together so as to attain improved combat readiness. This includes the 'American Regimental System' designed explicitly to create an environment that fosters cohesion and commitment. The United States Joint Chiefs of Staff in 1982 publicly recognized that, from its inception, the management of the All Volunteer Force focussed on recruiting and retention without sufficient sensitivity to discipline, esprit, cohesiveness and readiness. They emphasize that the military profession is not just a job: 'In return for selfless service and personal sacrifices, military personnel expect and deserve a quality of life reasonably comparable to that enjoyed by the society they protect; a strong system of institutional support to fortify their dedication and loyalty and nourish the unit cohesiveness, personal readiness, and teamwork essential in battle; and a stable predictable pay system.'[22]

In Canada, a number of studies such as the Base Concept Study, the Unification Task Force Report, the Cotton and Gian Studies, the Unit Administrative Study and the Kellet Study on combat motivation (which incidentally has been published in book form) have led to significant evolutionary changes in doctrine and organizational climate. Changes in administration have reinforced the regimental system significantly; changes in the individual training system now allow some environmental training for recruits. Most significantly, 2 PPCLI (Princess Patricia's Canada Light Infantry) rotates to Germany as a unit for a four-year tour in June 1984.

The Canadian Forces have, however, been trying despite these changes to find a corporate 'ethos' which would apply to all 'trades.' In looking into the future, we must ask whether such an approach will continue to be possible. Will standard forces-wide training and organizational concepts ever meet the needs of each of the operational elements of the forces? Will they, at the same time, maintain a suitable environment for the highly specialized, non-combatant technicians who will make up the major part of the peacetime forces?

A solution that appears to be simply 'happening' may well be the most appropriate one. It has been described by a respected military sociologist as 'a plural format emerging with three functional areas discernible: the *combat arms* based on the regimental system and traditional values; the *sea operations*, with unique institutions and internal culture; the *administrative-technical support element* where the value of technical expertise holds sway ...'[23] I, myself, would like to expand that. Obviously, pilots have a most vocational orientation; then there are those technicians who have to go to sea and those who serve in the field in support of the army. However, I do support the broad thrust of his conclusions.

Legitimacy

The need for both defence and defence forces to be legitimate in the eyes of the nation is the last factor affecting recruiting and retention. It impacts on the complete range of proposed solutions and I will therefore conclude and summarize my presentation while discussing it.

Although we have not seriously considered conscripted standing forces, we have suggested that a suitable policy option based mainly on compulsion would most likely require four to six months of obligatory training so as to provide the sizeable well trained reserves that we will need. A somewhat less effective, albeit cheaper, variant of this approach would involve simply registering graduating classes for rapid call-up in case of emergency. Neither of these approaches, however, could even be seriously suggested, much less implemented, unless the Canadian population was in full support of the necessity for such measures.

This factor also impacts on options which depend upon voluntarism. A voluntary regular force can, if small enough, and provided with good pay and benefits, probably be recruited with little concern for public attitudes; but the funding for such a force depends on the public perception of the defence requirement.

More importantly, we have suggested that our most critical need will be to maintain large voluntary reserves, generated perhaps by short engagements in the operational sector of the Regular Force so as to build up fit and ready Supplementary Reserves; and on a voluntary full-time period of training (four to six months) to build up the Primary Reserves. Public legitimacy is essential if the incentives for such an approach are to include the necessary benefits, such as subsidized post-secondary training, preference in government hiring, or income tax advantages. More importantly, the maintenance of an effective voluntary reserve is impossible without a 'vocational' orientation within the operational sector of the forces and without the full support of the local community of which the reserves are a part. Neither an attractive benefits package nor the force of law are sufficient; reservists 'genuinely consider themselves to be true volunteers ... motivated more by an obligation to serve, public recognition, and the opportunity to work at a stimulating contrast to their full-time occupation.'[24] They require the support and approval of their friends, parents, teachers, pastor, employer, news media and politicians.

We have a long way to go to get this type of support. Last week, Charles Lynch implicitly underlined this situation when, on Remembrance Day, he wrote: '... the impression deepens that Canada in two world wars took part in dark and sinister acts against humanity ... (yet) without (those Canadians who fought) the wars might have gone to Germany and Italy and Japan ... and the squabbles that perplex us today would be of a different and more sombre kind.'[25] NATO's current strategy for maintaining peace and security is being subjected daily to ever more well orchestrated attack by many of those community leaders on whom the legitimacy of our regular and reserve forces is most dependant.

I have wandered, in this discussion, a long way from technology and demography. Technical systems and personnel management systems are useless if there is no national understanding of the rationale and need for defence. We should perhaps examine to what extent the communications techniques, including advocacy advertising, which have been used to pursue changes in public attitudes with regard to social issues, inflation and so forth and which make the federal government the largest advertiser in this country, could not also be used in explaining defence issues.

In the final analysis, however, security and the continued maintenance of the unprecedented 37 years of peace that we have had in the West is not something that can be left to the Department of National Defence or even to the government; it is the business of every Canadian and, in particular, every community leader. Canada's contribution to world peace, and its own security, is vitally dependent on their example and leadership. The Canadian Institute of Strategic Studies plays a vital role in this regard and it has been an honour for me to address you.

Notes

1 Henry A. Kissinger, 'Strategy and the Atlantic Alliance,' *Survival* (September/October 1982), 196 discusses this issue.

2 General Bernard W. Rogers, 'The Atlantic Alliance: Prescriptions for a Difficult Decade,' *Foreign Affairs* (summer 1982), 1152.

3 Senator Sam Nunn, 'Conventional Forces and Alliance Strategy,' *Survival* (September/October 1982), 234.

4 Adriaan Ploeg, *Manpower Issues for NATO*, NATO Military Committee International Secretariat, draft presented by Rapporteur May 1982, 31.

5 Sub-Committee on National Defence of the Standing Senate Committee on Foreign Affairs, *Manpower in Canada's Armed Forces* (Ministry of Supplies and Services Canada 1982), 18-19.

6 Richard Halloran, 'U.S. Would Expand Combat Stockpile,' *The New York Times* (22 Aug 1982).

7 Sub-Committee *op cit*, 56.

8 The Honourable J. Gilles Lamontagne, *Parliament and Defence Policy*, (Toronto, Canadian Institute of Strategic Studies, 1982), 109.

9 A.J. Barker, *The Yom Kippur War* (Ballantine 1974), 123.

10 Dr Shabtai Noy, 'Division-Based Psychiatry in Intensive War Situations,' *Journal of the Royal Army Medical Corps* (1982), 100.

11 United States Joint Chiefs of Staff, *United States Military Posture for FY 1983*, 52.

12 This calculation provided by the Director of Recruiting and Selection.

13 G.M. Rampton *et al*, 'Military Manpower Trends in Canada: Attrition, Widening the Manpower Base and Other Perspectives.' Paper prepared for a meeting of TTCP(U) Technical Panel UTP-3, Military Manpower Trends, Cambridge, England, October 1978.
14 J.L. Granatstein quoted in W.A. Morrison, *The Voice of Defence* (DND, Canada, 1982), 194.
15 George F. Stanley, *Canada's Soldiers* (Macmillan, 1960), 213.
16 Comptroller General of the United States, *Report to Congress: Manpower Effectiveness of the All-Volunteer Force* (United States General Accounting Office, 1981).
17 Ploeg *op cit*, 4.
18 *Ibid*, 26.
19 Charles C. Moskos, Jr, 'From Institution to Occupation,' *Armed Forces and Society*, vol 4, no 1 (fall 1977).
20 Franklin D. Margiotta, 'Changing Military Manpower Realities: Implications for the Next Decade,' *Evolving Strategic Realities: Implications for United States Policy-Makers*, 1980, 115.
21 Richard A. Gabriel, *et al, Crisis in Command* (New York: Hill and Wang, 1978), 50.
22 United States Joint Chiefs *op cit*, 56.
23 Cotton, *et al*, 'Canada's Professional Military; the limits of civilianization,' *Armed Forces and Society*, vol 4, no 3 (spring 1978), 385.
24 BGen P.J. Mitchell, 'Understanding the Militia as a Precondition of Supporting and Fostering it,' *Canadian Defence Quarterly*, vol 8, nos 1 and 2 (1978).
25 Charles Lynch, 'Time to Recall Days when Canada Stood Tall,' *Citizen* (11 November 1982).
26 'Best We Forget?' *The Toronto Sun* (11 November 1982).

Brian MacDonald

Thank you very much, Colonel Liston. That was a superb presentation.

W.J. Yost, Emergency Planning Canada

I think that we are all in agreement with everything you say, Colonel Liston.

I want to ask you about the total preoccupation with the people in uniform when we are talking about changes in technology that are affecting the forces. We should be zeroing in on a total defence structure with civilians integrated into the high tech areas. We already have this in aircraft support, for example, where all of our third line is now in civilian hands. This is the way it is going in other service areas. The technicians in the Argentine that we heard about were not there by accident. They were actually required by those armed forces. I think you will see that happening more and more. The American general whom you quoted talking about his echelons behind the front line was describing this very well. Many of those echelons, as they got further back, should be an integrated military-civilian group – and are, in fact.

What is your view of this total defence approach to national defence?

Liston

Yes. When I noted the dichotomy between the occupational approach and the vocational approach, I meant that in the operational sector – the fighting pilots, the ships at sea, the chief elements of the armies – we must have a very vocational approach. But on the support side, the technical side – the element behind armies – there will be a much greater integration. There will be factory teams accompanying forces into the field, and conversely there will be military support personnel in the factories. We are currently taking an approach where high tech has allowed our skilled technicians to do more maintenance than they have been doing in the past. But if we were in a situation where we could not maintain that high technical competence, then we would be much more dependent on factories and factory teams than we are proposing to be at the moment.

In summary, I support your comments completely.

LCol W.R.C. Little, CISS

Mr Chairman, I would like to make a comment.

I would like to suggest that the priorities presented this morning by Colonel Liston are reversed. I think our major priority must be to have our political leaders and our government behind our activities.

I speak to groups of parents and teachers in my community to try to get them to understand that there is something honourable, something necessary, something positive in encouraging their children (and in the case of school teachers, their students) to join the local militia unit of their choice. In my area there are three of them; all three service environments are represented. But we are still suffering from the thing I call the Vietnam syndrome. The people I talk to are a cross-section of Canadian citizens from every facet of our social structure. They do not trust government when it comes to talking about the necessity for armed forces. Having said that, I would recommend to George Bell, who has led us so well over the years, that we do not stop trying to convince our government to regain the trust of the people and to point out the requirement for manpower to support a Canadian national defence policy.

Brian MacDonald

Thank you, Colonel Little. We can assure you that the CISS, under the leadership of Dr Bell, will continue to do as much as we have done in the past, and will continue to pursue those objects with ever increasing vigour in the future.

LCol J.A. Cameron

Colonel Liston, there is one basic difference between the Canadian situation and the American as far as reserve forces are concerned. The reserve forces of the

United States are a creature of the Congress and are protected by an act of Congress. If an individual voluntarily decides to serve his country, he does not have to fear for his job or his promotion. We do not have that support. Until there is that support, right from the very top of our government and political system, our reserve forces will keep scratching as they have for many years. One thing that must change in this country, and this touches on what Colonel Little said, is that the government must very visibly show itself to be in support of the forces, particularly the reserves.

Liston

I feel that employers must be able to protect reservists without losing money. They need tax write-offs so that people who have been let go may be temporarily replaced; there must also be incentives for them to encourage their people to go. I would view that as a necessity, just to maintain the current level of the reserves. However, the main thrust of my speech is not how to keep the reserves at their current level. The main thrust is on how the reserves can meet the operational commitments which I proposed that they should be able to undertake. These are far beyond their current capability.

Major John Hasek

The business of communication is the most important part of all. Many people do not realize how extraordinarily difficult it is to communicate from the inside to the world at large. Part of the problem for those who are still in the armed services is that every time one speaks it is viewed as the official word from the government. In this way, professional officers get confused with civil servants. In an earlier meeting here the deputy minister, Mr Nixon, gave us his view of what the strategy and policy of the Canadian Armed Forces should be. That version, of course, reflected totally the government's political view at the time. But there has been no place where the professional officer could give his views, which are politically untainted. There must be some way we can speak without it being interpreted as a knock at the government. Unless we can do that, we will have to wait until the public at large somehow becomes attuned to our views. It is a very difficult question, as those of you who have served around National Defence headquarters and who have achieved higher rank must know.

Liston

There are two parts to this issue. One is whether the military can speak out against the government. I think we are deluding ourselves in thinking that this will ever be allowed – and I'm not sure that it should.

On the other hand, there is the question of whether the government should publicly support its own position. I made a very tentative suggestion that we should have a more active approach to reflecting our own policy. Whether you agree with the policy or not is another issue. I suggest that we communicate the policy that we have in a more active and aggressive fashion.

JOHN FULTON

THE IMPACT OF HIGH TECH ON ARMED FORCES ORGANIZATION DESIGN

Brian MacDonald

It is now my pleasure to call upon Colonel Cameron to introduce our next speaker.

LCol J.A. Cameron

Ladies and gentlemen: It has been fascinating to me to have heard of the lessons learned and the lessons unlearned in the Falklands affair. One lesson that I learned a long time ago is that unless events are analyzed against a broad background of experience, conclusions drawn can have very embarrassing results.

Our next speaker has a most suitable background to take those lessons and to draw the right conclusions from them. LCol John Fulton, who is with the United States Army Training and Doctrine Command, had two tours in Vietnam as an infantry platoon commander, so he probably appreciates some of the problems we have been discussing at the grass-roots level. He has also just finished two assignments in Germany where he was the executive officer of an infantry battalion and the operations officer of an infantry brigade. He is a graduate of the University of Washington in Seattle, and he is also a graduate of Command and Staff Course at Fort Leavenworth. His current assignment is with the Deputy Chief of Staff of Doctrine, Headquarters Training and Doctrine Command, at Fort Munroe.

LCol John Fulton

Thank you, Colonel Cameron.

On behalf of the United States Army in general, and of my commander, General Glen Otis, at Training and Doctrine Command at Fort Munroe, I am very pleased to come and speak to you today about our future war-fighting concept. It is rather broad and general in nature. I speak to you primarily as a soldier who represents many other combat officers in staff jobs who have been in the field. We were brought into headquarters and asked to think about the best way to fight – to look at the nature of future warfare, and to look at forces in the field. I cannot speak to you in relation to the political ideas or grand strategy that support this, although we are aware of those.

I would have liked to have been here yesterday to hear the discussion of the Falklands war. We have not yet at Headquarters Training and Doctrine Command received an extensive and detailed briefing on that conflict. We pretend to know about it and we have a general impression of it from which a lot of conclusions have been drawn. But we have not yet gone into the details.

In answer to Colonel Little's remark about the Vietnam syndrome – it is very much with us still. Last week I attended with my family the dedication of the Vietnam memorial in Washington, DC. I took my ten-year-old daughter and I hope that some time in the future she will understand the significance of that war, but I don't think my generation will ever grasp it.

The Air/Land Battle 2000 concept is described in what might best be described as a White Paper. It is not a field manual. It does not represent current doctrine, since it represents a future vision.

First of all, let me bring to your attention the difference between a current doctrine and a future concept. We have a current doctrine now promulgated in one of our principle war-fighting manuals – Field Manual 100-5. We call it Air/Land Battle, and no doubt some of you are familiar with it. It is described by General Rogers, as a more aggressively oriented war-fighting tactic. That's a description of what we think that doctrine represents, as opposed to a future concept – the Air/Land Battle 2000 – a totally different concept. A lot of people mix the two together. They see the words 'air/land' and they think it is one entity. It is not. It is two separate entities. The doctrine described in the manual, if we had to go to war tomorrow, would be the basic tactic that we would use. The new concept does not have a tactic yet attached to it. It is a total force picture, dealing with the period 1995 to 2015. That is the concept I am going to talk to you about.

The American Army has been undergoing a lot of serious changes. One of them is the modernization of a lot of equipment: the M-1 tank, new fighting vehicles –

those are just the tip of the iceberg. Beneath that are somewhere between 100 and 120 other systems that are coming into the inventory – new families of radios and vehicles. With that kind of an equipment renaissance taking place, we suddenly realized we had to change the doctrines. So we began to examine what we had as a previous doctrine, the Active Defence, and we came up with the Air/Land Battle.

We feel now, with the Air/Land Battle 2000, that we have a much clearer picture. We have set down in a single document a vision of the future for the armed forces – for the Army especially, but it has implications for the other services. We are trying to build toward that.

The Active Defence was the previous doctrine that was adopted for the field in 1976. It was a major doctrinal change that the Army made after the Vietnam war. Four other ideas – the Central Battle, the Second Echelon, the Integrated Battlefield, and the Extended Battlefield – were terms that many of you have read about or heard mentioned in briefings, but they were never major doctrinal changes of direction. The Active Defence was still there as the basic tactic. The others could be considered as refinements or new ideas that were eventually incorporated into the Air/Land Battle doctrine that has just been published in 1982. The United States Army, in the six years from 1976 to 1982, made that single doctrinal change, although you can see that it is an evolutionary process and it will continue to be an evolutionary process in the future.

Let me move now into the trends. I'll only talk about a few of them. First, let me deal with Soviet power projection into the 21st century, in order to get a threat assessment that is more than just a weapons and technology summary. We are still looking at the Soviet Union as a future threat in the 21st century – not necessarily trying to gain hegemony in the world as we thought it was in the 1950s, but stirring up turmoil, discrediting the West, discrediting the United States and the United Nations, and exploiting Third World countries. We see a diffusion of potential warfare, a focus away from a central battle in Europe into more areas of the world.

The study assessed the strategic material depending on the United States and noted the increasing percentage of imports of strategic materials on which not only our industrial base in the United States depends, but also our military-industrial base. There are several implications. We either have to move toward synthetics, metal matrices and composites, or develop a strategy that preserves or continues to provide access to those original sources, or motivate the industrial base to get at other hard sources of these minerals, such as bottom-of-the-ocean mining, or possibly stockpiling.

The Air/Land Battle 2000 represents our future concept of warfare. Between now and its implementation, there will probably be another four or five different major fighting ideas that will fall in between those two major concepts, the Air/Land Battle and the Air/Land Battle 2000. The Army feels that six years is an appropriate gap for changes in fighting tactics.

Let me move into the concept itself. We begin with a quotation from Napoleon.

> With an army inferior in numbers, in cavalry and artillery, a commander must avoid a general action. He must make up the deficiency in numbers by rapidity of movements; want of artillery, by the nature of his manoeuvres; inferiority of cavalry, by the choice of positions. In such circumstances the morale of the soldier is a great factor. —Napoleon, Maxim No. 10.

The quotation reminds us that there are some basic underlying fundamentals and principles of warfare. The new concepts do not change that. The basic body of knowledge about how ground forces and naval forces supported by air operate is still there. All we have done is rearrange it into a different package, a different tactic. The changes are not revolutionary, but evolutionary, and are based on fundamental principles.

Another reason a quotation from Napoleon is appropriate is that we see the nature of warfare changing as we enter the 21st century, much as it changed around the beginning of the 19th century with the initiation of the model of the total war at the time of the Napoleonic Wars. As we approach the 21st century, we may be turning slightly away from that total war concept.

The methodology used to develop the new concept is based on three elements: a trends assessment, a future environment assessment, and a set of future implications for the military.

The following trends were considered significant and were examined in detail:

- Proliferation of nuclear technology
- Increased foreign investment in technology
- Dwindling United States production base
- Decreasing United States and Soviet military age population
- Increasing Third World populations
- Growing world-wide urbanization
- More diverse lifestyles
- United States becoming information-based society
- Soviet power projection
- United States and Soviet investment imbalance
- World political and economic interdependence
- More issues and less consensus
- Decline in earth's capacity to sustain its population
- Energy dependency
- Strategic materiel dependency
- Proliferation of arms

Now let me talk about another trend - world political and economic interdependence. As you know, we have a problem with our declining industrial base. For example, when the Ford Escort came out, it was advertised as a real world car. Everybody thought that was a joke. How could the good old Ford family wagon be a 'real world car'? But when you look at it closely you find out that many of the parts and components were manufactured in other countries (Japan, Spain, Brazil, Britain). So we may be subject to some technological as well as material interdiction with this trend. We analyzed some of our prized military equipment: for example, the M-1 tank. We found that it is pretty much an all-American product, but not necessarily all United States. Parts of the rolled steel on the front end and one of the gun computers are manufactured right here in Canada. That doesn't worry us, by the way!

Now let me talk about still another trend: more issues and less consensus. This is a sociological trend in which it is becoming more and more difficult to motivate our nation toward a single goal. We have a proliferation of issues, perhaps as a result of the mass media and the tremendous mobility of our population. We are moving away from the consensus we had after Pearl Harbor 1941, when the entire nation went out and enlisted. We have had a very difficult time in the last couple of months just getting young men to register for the draft. The concept itself tries to address that problem. Since the armed forces of the United States will never get a clear direction from our national government, from Congress or from national leadership, the services, and the Army in particular, will have to take some sort of future view. We will have to determine our future organizational direction for ourselves. I'm not preaching a coup here, or taking over the government. We will have to determine our direction and try to communicate that viewpoint to our national leadership. This concept does that and it has a high level of recognition within both the Congress and the leadership of the government. It has been briefed as high as the Vice President of the United States.

On the trend toward the proliferation of nuclear technology, it was estimated in the study that by 1981 there would be six nations world-wide that would have a nuclear weapons capability (I think there are only five), 31 by 1990, and an unpredictable number by 2000 because there are too many other variants in this trend. This proliferation tells us in the general picture that there is a possibility for smaller nation nuclear exchange. Whereas we had been looking at a world model where the Soviet Union and the United States were the principal nuclear combatants, homeland to homeland, we could have a situation arising where two smaller nations completely outside the spheres of influence of both of those superpowers engage in some kind of a nuclear contest. That is a situation we don't like to think about and one with which we are not at this time prepared to cope.

A demographic picture of the United States shows that we have the same problem that you have in Canada – a declining population in the age group 18 to 24, that slice of young men we normally expect to provide manpower for our services. The implications here are several. You can move toward a less manpower intensive force with more high technology: for example, move towards weapons crews instead of an infantry squad. It may mean that we have to retain older people on active duty and revise our officer and senior NCO retention policy, make better use of women, and continue to push to reinstate the draft – which *is* a tradition in the United States, although Vietnam has seriously shaken it.

The trend of growing world-wide urbanization is one that has more direct military application. When we look at the central battle as it would occur in Europe, we see large cities and urban sprawl. At one time it was a four- to six-hour drive to circumvent Frankfurt with a battalion or a brigade. Now it is difficult to get around within two or three days. Traditionally military forces have tried to avoid urban areas, but by the year 2000 you may not be able to distinguish between Frankfurt and Heidelberg. The implication here is that we must move toward forces that can cope with urban terrain.

In general, what these trends seem to tell us is that there is an East-West political division, but military confrontation has been further complicated by a North-South social and economic confrontation.

It is possible, then, to draw the following set of implications for the army:

- It must be prepared to fight anywhere;
- It must win the land battle;
- It must maintain weapons parity;
- It must avoid high combat losses;
- Initial battles are critical;
- It must develop strategic mobility.

First of all, we must be prepared to fight anywhere. When we went into Vietnam we had a lot of problems. It was a long ten-year war, but some things we never did develop. For example, we did not develop a good doctrine, equipment or techniques for tunnel fighting. Jungle boots and uniforms were issued late in the conflict. Even such things as rope ladders to reach down through the jungle canopy and insert small teams were not available. We had the helicopters up above and the forces on the ground but we couldn't extract the wounded. Many of these things must be thought about and developed prior to the conflict. We have looked at World War II and Central Europe as a World War III analogue in the focus and planning of most of our force development, whether it be material or doctrine. Now we are beginning to open up the focus. We must not lose track of the fact that the central battle is very important, but we must look at contingency areas in other parts of the world and prepare for them. The Rapid Deployment Joint Task Force is a move in that direction.

The second point: we must win the land battle. This is a theme that is carried on from the Air/Land Battle, from our current doctrine. It simply expresses a more offensive capability, clearer identification of the objectives, and a recognition that the Soviet Union is primarily a land-based, land-power nation.

With respect to weapons parity of combatants, we cannot match the Soviet Union man for man, weapons system for weapons system, tank for tank, airplane for airplane. The quotation from Napoleon suggests that one has to come up with some sort of indirect approach, a reliance on quality of weapons, reliance on technology, tactics, leadership, and national will.

We must avoid high combat losses. When warfare intensity goes up there will be more casualties. This presents a tremendous burden for the medical support facilities on the battlefield and indicates also to us that when we talk about manoeuvre warfare, which this new concept tries to portray, we find that it is traditionally more intense and more lethal than attrition-defensive warfare. We are trying to move away slightly from the active defence role that we saw ourselves playing.

Next, the criticality of initial battles. Some of our doctrines used to talk about winning the first battle. This is not the same thing. In World War II initially we underwent some losses prior to mobilization and gearing up the industrial bases, drafting and deployment of forces. We went down hill. After gearing up, we went on to achieve a complete defeat of the enemy - with our allies, of course. In future wars whoever is on the ascendancy from the start may be more of a determining factor than the total military might of the nations who are fighting. The impression that we get from the Falklands war is that the total military mights were not as important as what went on in the theatre. In terms of a future European conflict, it is unsettling to note that that initial deficiency could be felt in a battle only kilometres from the East-West German border. That may be the biggest lesson from this.

We must have strategic mobility. We must be able to fight in more places at one time. The services in general have sought to develop a much more unified effort, whether it be in loading plans for equipment, construction of aircraft to fit our military equipment, or packaging forces in a fashion so that they do better than they did in Iran where command and control was at such a high level that decisions in the field couldn't be influenced.

The operational concept itself flows from the trend and environmental analysis. It develops into what is going to look like a tactic, a change in the way that we fight, and we try to capture it in something we call The Essence.

ESSENCE OF THE AIR/LAND BATTLE 2000
OPERATIONAL CONCEPT

A style of waging war in which agility, deception, manoeuvre, firepower and all other tools of combat are used to face the enemy with a succession of dangerous and unexpected situations more rapidly than he can react to them.

It is simply a short statement that describes how we feel future warfare should be conducted. It is more a change in degree rather than a totally different approach. It simply recognizes what I think some people refer to as the indirect approach. You can't destroy the enemy in detail. You must rely on manoeuvre and deception and other battlefield characteristics.

In order to explain it, we use an analogy. In our Active Defence, the way we were planning for a central battle in Europe, and even to a degree in our Air/Land Battle, the war looks something like a football game. Heavy forces are arrayed along a line of contact, there is sporadic contact tightly controlled by a play or a plan of action and then a separation in which you plan another play, there is a distinct orientation between offensive and defensive modes of warfare, and there is very tightly controlled command and control in which the players, like our units, tend to have specific instructions or missions to carry out. A lot of our orders in the field are written in terms that you might use to describe a football game. It's all very terrain oriented.

We see our doctrine, with the trends toward high lethality and mobility, taking us to something more like soccer or ice hockey. Here you have a more open game. There is a general orientation of forces against one another, but there is a distinct intermingling. Forces are not on a line. There is less distinction between offensive and defensive modes of operation. Commanders and units play to an overall strategy with no time outs. It is continuous play, and they do things based on their positioning and the overall strategy. In that type of warfare, the units do not have detailed plans.

Let me drop down now to another level to look specifically at the battlefield characteristics of this Air/Land Battle as it would take place. We find large quantities of sophisticated combat systems, and a battle expanded into air space and in depth of enemy formations to a possibility of 300 kilometres. With one exception, these are not great changes in kind; they are changes in degree or intensity and emphasis. The exception is that chemical-nuclear-electronic warfare has not been engaged in to a great extent.

In our current Air/Land Battle doctrine we propose that the corps commander have the ability to conduct reconnaissance surveillance and be able to shoot and destroy the enemy as much as 72 hours forward of his front line positions. That translates in some terrain to as much as 100 to 150 kilometres. If that is so, and if we can give him that capability now or in the near future, it would certainly be a battle that could extend to 300 kilometres forward, rearward or in any direction in the far future.

We have a series of principles that we have associated with this concept. They do not replace the traditional set of principles of war, which I am sure you all endorse in some fashion yourselves - things like mass, manoeuvre, economy of force, security, and surprise. They are simply inherent to the concept, though they may have an influence on some initiative later on to rewrite or re-evaluate the current nine principles of war. They looked so good that we dropped them into current doctrine. They are the following:

- Agility
- Initiative
- Depth
- Time
- Synchronization

The Principle of Initiative simply tries to capture that spread out, independent operation that is going to be required. A lot of people interpret that to mean a beehive type of tactics, completely out of control, with commanders doing what they want, selecting objectives on their own without regard to an overall scheme. We do not mean that. We feel that commanders must be well grounded in basic tactics, but have more leeway than they have right now to conduct their operation and to take more risk on the battlefield.

The Germans have a term they call *austrog taktik* to describe this mode of warfare. It means roughly operating within the scheme of manoeuvre of the general strategy of the higher commander. We don't use that term, but we are trying to get at the same thing.

Depth on the battlefield simply describes this spread-out nature of the battlefield, both forward and rearward. In our current Air/Land Battle doctrine we talk about the deep battle against the second echelon, the assault battle with his forward echelons, and the possibility of having to conduct a rear battle if he does an envelopment or penetration through our lines. This principle of depth simply carries that scheme farther. It indicates that we must be prepared to fight in more directions at one time.

The Principle of Agility is somewhat like the principle of manoeuvre. We want to move toward more manoeuvre warfare, which is not there in Active Defence because we didn't have the systems to support it. When we came out of Vietnam in 1972 and looked at Central Europe, we found that our forces there needed a lot of repair. Probably the Active Defence, which was devised sometime around 1974 and published in 1976 was the best we could do based on the current systems in the field at that time. We feel that now new equipment has given us the ability to do more. This carries forward the theme that says you must win by going on the offensive and by manoeuvring.

The Principle of Time may be a further refinement of the Principle of Mass. What may have occurred in three days at the Battle of Gettysburg in 1863 may occur within 20 minutes in the battle of the future. The intensity spread over that future battlefield could equal the three-day Battle of Gettysburg. This is the area where we see the greatest emphasis on high technology – giving the commander those aids that he needs to be able to make decisions, manoeuvre his forces and conduct the battle, whether by graphic displays and maps, or refined intelligence summaries. They will replace the long printout lists of grid co-ordinates that the commander now hands over to somebody who takes three hours to plot them on a map, by which time the battle has progressed and the information is obsolete. There is a lot of man-machine interface involved in this, whether it be electric pencils on the map that assist him in planning the movement of units, or instrumentation in the cockpit of aircraft or other combat vehicles. This is where the Jules Verne image of future warfare comes into play more than anywhere else.

By Synchronization we mean not an elaborate scheme of manoeuvre or an elaborate timetable for the movement of units, but just a blending together of many combat systems with good doctrine to back it up. It takes the new equipment and doctrine and puts it all together in a package that can be used in an efficient way by combat forces.

Right now the American Army has a lot of things that are highly specialized. The best example is in engineering. We have specialized bridging equipment located in units piled up some place 70 kilometres in the rear. When a combat unit runs into an obstacle, like a river, it has to call the bridging unit to provide a bridge. We'd like to see that gap-crossing capability built into the force, whether it be in the swimability of vehicles or expedient systems carried right with the force. Our engineering laboratories are working with some technology using foam. They claim they can 'foam' a bridge by stringing a couple of large cable wires across a moving body of water, and then rolling out tubes that dispense foam and put down a pontoon-like styrafoam over which you can drive vehicles. That's on the drawing board, not in a battlefield configuration.

The Air/Land Battle 2000 concept recognizes the necessity for things like that and provides direction to our development community. It establishes some priorities for them on this type of high technology. We have found in the past that our development community, Readiness Command, which builds and develops all our equipment, did not have a clear message from the user on the combat side of the house as to what we needed. This concept has been briefed to them and they feel they now have a much more coherent picture of the future that provides them with a lot of direction in their tech base, their experimentation, and even provides some direction to the civilian industrial community.

I would like to pick up on two of the things that Colonel Liston talked about. First, the man-machine interface, which is built into the equipment. The laboratories developing the future helicopter in the aviation community are probably as far ahead as anybody on this. They are talking about such systems as automatic evasion devices in aircraft, unless the pilot overrides the system; and automatic prioritization of ground targets and engagement of ground targets with weapons systems - again unless the pilot comes in and takes over the controls. That is very innovative.

A second is the training aspect. The kids all play computer games, video games. We recognize that that is a subset of this man-machine interface. If you watch the kids play these things, you find that they don't read the instructions like you and I would. They simply get a visual picture, based on the way the machine is set up, and whether it's a lever or a button or a flipper, they know how to play it. The message comes through to them in a different language. Somehow that has to be built into our equipment; that's the type of man-machine interface that has to be considered. However, I wouldn't go so far as to suggest that you have a video game display in a tank.

Many people ask what we are talking about in terms of organizational structures. Is the American Army going to do away with divisions? Are we talking about a follow-on to Division 86?

We will probably still have our battalions and brigades and our old regimental system. The current lineage of distinguished units will remain in the army. I think that is necessary. But what we are trying to do is to devise a force which will take some of these rather cumbersome headquarters that we have in the field and streamline them, to give them more manoeuvrability so they can conduct this manoeuvre warfare. They can't do it right now. Our divisions only have three or four brigades. They don't have enough units to manoeuvre, so they are continually going through task reorganizations in the field in order to be able to achieve manoeuvre bodies.

The *Close Combat Force* is the primary tactical combat organization. It represents what might be called a separate brigade or a small division of about four to five thousand, highly manoeuvrable and as lethal as a division. The next level up, the *Land Battle Force*, is in the position where we now have our divisions. It would be a command and control headquarters only, somewhat like the United States brigade without the division support command and the division artillery. The next level is where the corps is: the *Air/Land Force* headquarters. The area support required for this type of battle, when you spread the units all out, is best accumulated we think now at a higher level. The theatre command or combined headquarters won't change. That will be a joint coalitional headquarters.

The Air/Land Force, the corps equivalent, is not an army corps any more, but something of a joint corps, with significant participation by the air force, and possibly navy and marine forces. Right now our corps has small liaison teams to the other services; we fight the land battle as an army show only. If the future is truly going to be an Air/Land concept we have to bring in the other services. We have a lot of joint working agreements right now with them. That's one of the stickier issues in the implementation of Air/Land Battle – how you put together those two force components in the field.

The forces controlled by that Air/Land Force Headquarters will look a bit like a mini Rapid Joint Deployment Task Force (RJDTF), in which you may have under a corps headquarters some army components (brigade-size or division-size forces), a marine component and even a sea component of some sort, plus an air wing.

The following is a list of force characteristics:

- Small self-sufficient organizations
- Highly mobile
- Firepower intensive
- Less manpower reliant
- Extremely agile – dispersion – mass capability
- Capable of fluid, continuous operations
- Rapid strategic deployability
- Extensive use of robotics
- Micro-electronic technology
- Electronic Warfare (EW) incorporated in all combat forces
- EW to destroy, disrupt, deceive
- Dedicated all-source intelligence
- Real time theatre intelligence
- Mobility – countermobility - survivability built in
- Information processing for logistics
- Less fuel consuming - energy efficient
- Invisible equipment (multiple or stand-off signatures)
- Knowledge of position
- Redundancy in command and control
- Reconstitution capability
- Less turbulent, cohesive (regimental system)
- Design for war – augmented for peace
- Family of vehicles
- Information pushed to commanders

The list is a shopping list of what has to be done in order to implement this concept. It is evolutionary. We don't see all of this in place in 1995. We will trend out to it at the same rate as this concept.

In 1995, we will still have the M-1 tank; we may even have some of the old M-60s. These force characteristics represent all four development areas: not just high technology and equipment; but also doctrinal changes; and organizational, training, and leadership initiatives as well.

Up to now, electronic warfare has consisted of jamming, deception and interception, but there is a significant potential in the area of laser and directed particle beam for development into destructive weapons. Right now, to put a particle beam into effect to destroy something at any distance takes a generator the size of a locomotive, but, based on trends in other areas, by the year 2000 that will be significantly reduced and we may end up with good, packageable electronic weapons on the future battlefield. It is somewhat like the development of gunpowder. When gunpowder first came on the battlefield it was not very effective but had a significant psychological impact. It took a long time before its effectiveness equalled that of the long bow.

By invisible equipment, what we mean is stand-off signatures or multiple signatures - not truly invisible.

Wars may be fought with weapons, but they are won by men. It is the spirit of men who follow and of the man who leads that gains the victory.

I guess Patton said it all. As a soldier that says to me that troops in the field are motivated by a clear-cut objective that they understand on their own terms and a war-winning strategy, again expressed at their level. All the beer and sodas and USO teams and morale support that we arrange does not make up for that. We hope that this new concept will give them what is required.

Sir, that completes the briefings. I would like to take any questions.

Brian MacDonald

Thank you very much, Colonel Fulton.

I will now ask Colonel Liston to join us on the platform.

OPEN FORUM

BGen Marcel Richard

Mr Chairman: A small statement to clarify another statement made yesterday about the Falklands. The Exocet is manufactured by the Société Nationale Industrielle Aerospatiale. The Super Etendard is manufactured by Avion Marcel Dassault Breguet. The Argentines had a contract, including third line, but it was not the manufacturers of Exocet who were in Argentina. I remind everybody that the President of France, François Mitterrand, was highly supportive of the British.

My question to Colonel Liston regards the dichotomy between high-tech specialists and low-tech slit trench diggers like myself. It's not an entirely new problem. In 1968, under the then-CDS General Allard, we approached the problem from the point of view of definition of tasks and fields of employment. We found that to stimulate recruiting and retention, by means of a compensation package with as much equity as possible, we could make use of the definition of tasks. My question is, does it exist? If it doesn't, is any thought being given to such a system in order to stimulate our high school graduates to get into less technical fields such as infantry or the combat arms?

Liston

What we have is the Land Operations Trade Reassignment Programme (LOTRP) which allows a soldier serving in the combat arms to remuster into a technical trade between his third and twelfth (I believe) years of service. There are 47 technical trades into which he can go. The programme has been successful from a recruiting point of view. It is viewed as an attractive come-on to the high school graduate who wants some marketable technical advancement. The receiver trades were initially reluctant; the common perception was one of illiterate infantrymen coming in to become electronic technicians. They have, however, become enthusiastic supporters of the programme. The only remaining reluctance is among some who are concerned about the effect it is going to have on the status of the combat arms when they see that the Armed Forces themselves are recognizing that a lot of young men will eventually want to leave their regiment and go into a technical trade. The proposal of 'pay fields' has been largely set aside in favour of 'equal pay for equal rank.'

Leonard Bertin

My question relates to finding substitutes for Certificate A, that pre-World War II certificate which provided us with so many of our officers ready-made in wartime and also for the OTCs that no longer exist. A senior officer in the OPP stated that they are now deriving a lot of their high quality recruits from 23 community colleges in Ontario that produce 3000 students in police work. They call it 'criminology' to appeal to current attitudes. We must look for something that really appeals to these young people. When they train people to run the Pickering nuclear power station, they don't risk the possibility of a billion dollar meltdown. They get CAE in Montreal to produce for them a simulator for that power station. I wonder why we are not spending very small amounts of money on providing simulators in our militia headquarters to teach tank driving or missile operation so that we have something that will appeal to these very intellectually stimulated young people. Now when they get there, there aren't many people around, there's not much to do, and they often go home because there is nothing that excites their intellects.

Liston

I think the Forces would agree with you completely in principle. One of the thrusts of Training Development is to develop these computers and simulators. In some of the more technical trades, we are going ahead at full speed. The problem is that the initial funding for this approach is terribly expensive, even though the benefits are substantial. Not only do you have to buy the equipment; it must also be programmed. It should be done, but when you line up all the other things that have to be done, it just doesn't rate high enough in the priorities.

Major John Hasek

Colonel Fulton: I don't think you can wait for the next generation to learn your lessons from Vietnam. The lesson that I learned in Vietnam, although I was there in a more detached capacity than you were, was that it took me about two months to grasp that my prior perceptions about what had gone on were totally at odds with the facts. After you people left, the South Vietnamese army, which we had learned to despise as being useless and corrupt, was strong and in far better fettle than the North Vietnamese army, who were by that time really on their uppers. We had access to both sides, you will remember. NVA groups marked on the map as 'regiments' were, in fact, 50 starving conscripts.

What your army failed to do was to tell your people what had happened. War, especially low-intensity war, is to a large extent a matter of perceptions. You won the war materially, and professional American soldiers know that – but it is a war that your public thinks you didn't win because you still haven't let your people know that reality. In any future war, part of the enemy's attack will be to try to isolate your army from the community. Unless your professional officer realizes that part of his responsibility is somehow to communicate what he is doing, he will be cut off from that society and you will end up with the same problem again.

Fulton

I think that is a fair assessment. Vietnam is undergoing extensive evaluation. What I just showed you is the culmination of that evaluation. Probably the biggest error we made was in not clearly identifying the objective. The book *Strategy in Vietnam*, written by a military officer, Harry Summers, analyzes the war in terms of the Principles of War. Your assessment is correct. We beat them in the field. We destroyed them at the tactical level. I don't think we ever had a large unit resolution. But strategically, we did lose the war, even though we won the battles.

Colonel Motiuk

Colonel Liston, I think you slanted your paper towards getting technicians. I think there is another issue. If you are going for an increase in the reserve forces, you have left out the officer corps. Who is going to officer this organization? We did away with our officer training corps in about the mid-1960s. Are we going to resurrect it? What sort of scheme would you suggest to staff the reserves with good officers, both technical and operational?

Liston

I am less concerned about recruiting technical people. They are available and in view of the challenge we offer them, we get them in quite easily. The problem I addressed concerns getting the youngsters that have this technical capacity, now some 60 per cent of our population, into the combat arms.

At the officer level, I believe the traditional COTC option is the best one, but I have not given this issue the same attention. I believe that we could get the officers through COTC, although we might have to offer more benefits than in the past.

BGen C.L. Kirby

I'd like to say that for the first 20 years of my service, the high command of the army and their political masters regarded the Canadian war effort through the narrow end of the telescope. For the last 15 years, they have regarded it through the fat end of the telescope and therefore as something very far away and very small. Despite Colonel Liston's paper, I can see from many of the remarks made that this sort of perception of our war effort continues.

When will the Canadian Forces stop regarding war in the abstract and start producing something more practical to guide our efforts? We inevitably these days talk about our peacetime commitments to NATO as if they represented our total capability in war. In the last war we produced almost a million people in our armed forces with half the population we have now. It strikes me that winning a long war must be our aim. We must look at our ultimate capabilities first and then work down all the factors, including high tech, to what we could support in war. I would suggest that a population of one million can support one division in action and the air forces and the navies and the production required. If that is invalid, I would like to see it proved invalid scientifically. The highest figure Colonel Liston mentioned was 500 000, about half what we produced with half the population in the last war. If we could separate our war capabilities from our peacetime commitments, peacetime commitments of a government not inclined to defend the country seriously, then it strikes me that we would be on the right track.

Further, I think that if you want to start on our educated youth and reintroduce them to the forces, the place to start is the COTC, for operational and institutional reasons as well as the practical possibilities. I feel that would be a signal to the country that the government was serious, and it seems to me very easy to accomplish.

I have one suggestion for Colonel Fulton. It strikes me that if you are going to describe initiative as a military requirement these days in relation to the definition of initiative in the civil world, then you must qualify it by calling it disciplined initiative.

Liston

General Burns has done an excellent calculation and has given us a formula for our total capability in his book *Manpower in the Canadian Army*. I think he concluded that Canada could put into the field 45 per cent of the cohort of 18- to 45-year-olds, the other 55 per cent being required at home. That, of course, assumes optimal utilization of the manpower of the country. I must admit I don't remember the figures of what the divisional slice is.

I have not attempted to suggest a massive change in defence policy but merely to point to the effect that technology could have on our current commitments. Our current commitments are limited and I am suggesting that the desire to reduce the reliance on nuclear weapons and therefore a need for our current commitments to be more sustainable is going to call for larger conventional forces. I have not addressed whether we have an appropriate commitment to NATO or not. Perhaps it would be better for people who are not serving to address that one.

The Minister, when he was here at our last meeting, indicated that he intended to submit to the Cabinet his assessment of exactly where we stand and what we would require to make our forces sustainable. That is a public commitment and I think I should leave it to him to publicly announce what further developments have taken place.

Fulton

We are trying to improve our capacity for disciplined initiative in the training of junior officers.

Ever since I have been in the army, and I think before that, the thrust of officer training has been toward resource management. We have sent a lot of our junior officers off to business administration, operations research, and that type of advanced education to support their professional development. Now we are trying to get away from that a little bit, and to try to re-emphasize the idea that there is a doctrinal base. An officer cannot exercise any initiative at all if he doesn't know how to fight. If he is principally an administrator, a resource controller, then when he goes to fight (and we see this on a day-to-day basis) he can't train his troops, he can't manoeuvre and he can't talk on the radio.

Dr Friedman

Colonel Fulton, I find the Air/Land 2000 concept very exciting. I wonder to what extent the NATO allies have been briefed?

Fulton

We tried to take our Air/Land Battle doctrine to NATO and we found some difficulties, although some people have completely accepted it. Because the Air/Land Battle 2000 is a future concept, something you can plan for, it has been more accepted. It has been introduced into Panel 11, it has been briefed to the NATO military committee, and we have unilateral working agreements with the French, the United Kingdom, and the Germans. In each case, we have had to take the written concept and rework it to suit language changes. This results in a proliferation of written documents, but it is the same core idea. And we have introduced it into the ABCA talks as well. It has been the subject of many joint studies.

Dr George Bell

Colonel Fulton, my question is an extension of Dr Friedman's question.

Right now in NATO we are dealing with about a 600-mile front in Central Europe, a defensive posture, a concept of forward defence, which imposes some limitations in terms of manoeuvre, in terms of depth and the use of terrain and mobility. How do you see your concept affecting the NATO strategy? How do you deal with mobility within the territory when we are posed linearly over 600 miles, giving the initiative for an offensive to the potential enemy?

Fulton

This concept is not a tactical configuration yet. It is a general direction for the development of forces and it postulates more emphasis on manoeuvre. The way we fight in Central Europe, how we are positioned on the East-West German border, the forward defence idea – it may not apply to that at all. It will apply in certain aspects, but it may not change the basic tactic.

Bell

To go back to Dr Friedman's presentation yesterday, we have a political determination of strategy which affects the battlefield concept. When you deal with Germany, for example, you have to have a forward defence concept. You cannot use the mobility that you plan.

Fulton

As a military planner, our obligation is to come up with the best plan to defeat the enemy. The political implications are there, but as a sort of ceiling which we cannot go above. In the Air/Land Battle we devised what we think is the best way to defeat a heavy Soviet force.

Bell

I agree with the concept. I see problems in applying it to our NATO commitment.

Colonel Liston, in your paper you posed the need for increased manpower. But in dealing with the subject, have we really put the time and space factor into it correctly, or at least sufficiently? When the Deputy Secretary General of NATO was here, speaking at our NATO – Thirty Years seminar, we discussed something that is never discussed in Canadian circles – rapid reinforcement. He said that rapid reinforcement in NATO terms meant that we had to be able to raise the force quickly in peacetime in a period of rising tension, hold it ready to move, be able to move it on short notice into the North Flank or to reinforce the Central Front. Can we deal with force development postures that do not provide a capability to raise men in situations short of war, in peacetime configurations, and move them? We appear to lack the legislation, the organization, and the compulsion capacity.

Liston

I agree. There are questions of legislation. There are questions about when would be the appropriate time to call them out. There are financial aspects. There are issues of how to equip these forces. Are they going to have the same equipment as the Regular Force? There is the problem of where that equipment is going to be. Is it going to be in Canada or prepositioned overseas? There are a great many issues.

I concentrated on what I considered to be the most central personnel issue: the basic concept of what type of reserves we need and how we should get them. The immediate reinforcements to bring our standing forces up to strength and to provide replacements within the first week or two should be former regulars who have recently served one or two years in the Regular Force and who can, with appropriate legislation, be immediately called back. In the meantime the Primary Reserves are being called out, with men and officers who have been trained properly. My personal hypothesis is that this will have taken four to six months of full-time training in central schools to give them basic training, plus collective training that they would take in their units. The necessary legislation would have to exist if this were to be possible.

Richard

Colonel Fulton, would you provide to us a brief sketch of the new United States regimental system? Is it addressed solely to the infantry? Where are the recruits trained? Finally, do you envisage more regimental distinctions being allowed and regimental traditions being expanded?

Fulton

I wish I could give it to you. There have been some recent decisions made by the Chief of Staff and they are fielding some units right now. It is not confined to the infantry. As a matter of fact, the first units going into Germany are artillery units. It is based to a certain extent on the British regimental system and tries to take advantage of that cohesion and stability of personnel. Because it is under review at the moment, I really don't know enough to say more about it. It is a viable issue, still on the front burner. There are still some problems with it, particularly when dealing with forward deployed forces overseas. It is going to require rotations of units back and forth to Germany, which at that distance is a significant problem.

LCol Michael Stevenson

One of the lessons of the Falklands at the low end of the scale was how effective fit battalions are, such as the parachute regiments and the Royal Marine commandos, while further down, some of the Guards regiments were not fit. That was very noticeable, although they are after all regular force units who had been on duty in London. If you are going to fly 25 000 ready reserves over to Europe, and in the case of the United States forces I think the figure is about two and a quarter million eventually and some of those within 30 days, we must address the simple problem of soldiers being really fit to fight, to carry heavy loads, to get through the first five or six days of stress, being able to march and dig, march and dig. I don't think anybody considers this seriously in the peacetime army. But I believe it is just as important as the high tech stuff at the other end, and in the case of the Canadian army, if I had to go to Europe, digging would be the most important thing I could possibly do.

Liston

I agree with your sentiments. In our regiments, despite the technology, we take great pride in our physical fitness. As for the reserves, we would hope that they would be as good as they possibly could be. If I were a reserve CO I would try to find ways of motivating my people to make it a point of personal pride. From what I've seen, there are a lot of reservists who are more fit than the regular force people. One consolation is that the other side is also calling up masses of reservists. So perhaps after the initial engagements, reservists will decide the issue on both sides.

LCol J.A. Cameron

Colonel Liston, it appears to be accepted that land, air and sea reservists who are former regulars or individuals who have served for a period of active service can be rapidly incorporated into the services for near-immediate active duty or combat duty. This is being done in the United States, certainly in the air reserves where individuals with former active service play a role on a three-, four- or five-day per month basis in crewing, for instance, the tankers of the Strategic Air Command. These are individuals who have had two or three years with the Air Force and in civilian life are back into the air element, flying or navigating or engineering or acting as ground crew in a civilian capacity for one of the air lines. They simply change their uniform when they take up their 48-hour or 72-hour duty with their reserve unit. That's very easy. One can go from a computer in a civilian capacity to a computer in a military capacity, and if they have had basic military training it isn't too difficult to make that transition.

However, like General Richard and a few others, I am a poor bloody combat type – or was, years ago – and I find it very difficult to visualize in the Canadian context taking someone who has spent four or five years in the regulars and has then been attached to a reserve unit and expecting him to suddenly overnight be ready to do combat. The reserve units aren't even equipped the same as the regular units. I believe that this has been addressed in the United States where reserve units that are designated for active duty are equipped the same as regular units, so that even though they may not have the training opportunities of regulars on a five- or six-day week basis, they at least can do it on a two-days-a-week basis or a four-days-per-month basis. They are not suddenly faced with something that is new and different when they are mobilized.

Liston

There are two types of reserves. One is presently called the Supplementary

Reserve and I am suggesting that at least the operational part of it should be a ready reserve. This is not the militia; it is not the naval reserve; it is not formed of people who train once a week. They are people who might only have a mobilization exercise once a year, where they would come into the unit that they would fight with, show that they are still fit, and find out where they would have to go when mobilization occurs. They would get their year's benefit for showing up that one time. This is a very cheap form of reserve. There is no training required because they have just left the Regular Force and have the necessary skills. They are, perhaps, no different from a fellow who is posted out of an infantry battalion into a staff job or some other job away from field soldiering. Of course, as technology changes, their usefulness changes, and even if the technology doesn't change, their usefulness degrades as time goes on. The French army has massive reserves because there are very few exemptions from compulsory service, but they use five years as a guideline. A reservist who has been out for more than five years is no longer useful in their mind. So I would suggest that the ready reserve I am talking about consists of people who have retired from the Regular Force in the last four or five years. Our current Supplementary Reserve has people who retired a great many years ago.

Of course, the Primary Reserves must be trained collectively. But lacking conscription we have to have some voluntary period of full-time individual training for them to be ready and then it is up to the unit to ensure that their collective training is done.

Kirby

I think it is essential that not only the mobilization of reserve forces but the operational use of reserve forces has to be considered. If you consider the jobs to be done on the battlefield, the operational use of different categories and different qualities of mobilized forces can be quite systematically organized. As these mobilized forces develop in action, you can expect more and more from them.

The Germans, for example, in the First War, used exactly this system. They had different categories of units and different categories of mobilized formations. Those categories were given jobs that suited them, physically and from an equipment standpoint, and therefore suited them tactically. The more regular formations operated around the depth and the stability they provided. It applied to the German army in the Second World War too. The really salient fact about them, and about the better but still low category divisions that supported the high tech edge, was not how peculiar they were (although they were peculiar to us) but how well, almost without exception, they fought.

Liston

I think that on the theoretical level, General Kirby, I agree with you completely. Indeed, you are one of those who have given me my prejudices.

However, I reiterate. I am dealing with Canada's *current* level of commitments and the reserves necessary to meet them, rather than an extensive mobilization. To meet our current level of commitments I am suggesting four levels. The first is the ready reserve that would be immediately integrated into the standing units. The second is the primary reserve that would have to be at a high degree of readiness – those units tasked to support our NATO commitment within anywhere from two to six weeks of M day. The third are those units for the defence of Canada. They can perhaps be at a lower level of readiness, not because of training restrictions so much as simply because we are not going to be able to equip all the units instantaneously in the immediate future, even if these concepts were accepted. So they would probably have lesser equipment and be at a lower level of readiness. The fourth and lowest level that I have suggested are the vital-point guards, a home defence organization. I suggest these categories because I have restrained myself to looking at current commitments rather than a vast mobilization.

Whereas the Americans are looking at the Air/Land Battle 2000 and at striking into the rear of the enemy's echelons, the European countries are being encouraged to look at mobilizing their vast numbers of reservists with simple equipment so that they can create almost a people's war type of situation in their own rear area and thus increase the depth.

BGen C.E. Beattie

The challenge of high tech is not so much how we cope with the black boxes and handle the electronics, but how we view ourselves in meeting the challenge. A couple of terms came out this morning, such as troops viewing themselves in two categories – vocational and occupational. That suggests to me that high tech can insidiously destroy our traditional military values and sow the seeds for a force to fall apart when it does get into combat. If we get part of our armed forces thinking that they are prima donnas simply because they are technicians, and that they don't have to develop and perform as combat soldiers, sailors or airmen in a war, then high tech will certainly work against us.

Liston

The issue that you bring up is very difficult: the question of the vocational versus the occupational approach. I think that we have three choices.

The first choice is to say that in Canada's unified forces we are going to have only a vocational approach; everybody is going to be a soldier. And that means we are going to maintain a system whereby we bring people in at the bottom, give them a soldier-type basic training, and then lead them in lock-step through their technical training. We would do this with the F-18 electronics technicians and similar tradesmen as well as with the combat arms. To become a corporal, that electronics technician would have to go on a junior NCO course, do bayonet drill, and master all the other skills of leadership in combat. That's one option, but there are great disadvantages to it which are obvious to everybody.

A second option is to try to find a compromise, which everybody would follow; a type of training and approach to soldiering that meets the needs of both the technicians and the soldiers. I suggest that is what we are trying to do right now. In our recruit schools we have not only infantry sergeants, but we have cooks, mechanics, and aero-engine techs who are teaching our recruits general military training. Having infantrymen, however, being taught their 'trade' by cooks, is perhaps not the ideal situation! But how can you criticize the technical trades who say they don't want their people being brought up in a system where their NCOs are not represented? We are not going to succeed, particularly in a peacetime army, in imposing a vocational combat arms ethos on all the technicians that we have in the three services. If we insist on having just one way, we're going to end up with our combat arms soldiers being brought up as technicians!

Therefore, the third option is perhaps the best, within the framework of unification, we should look at the missions of the various parts of the forces and ask what are their requirements in terms of ethos, training and attitudes. In the technical trades it takes a long time to train a technician. I have suggested ways of avoiding that, by bringing them in laterally from technical colleges. But at the moment they have to go up the ladder, with many years as a private and another lengthy period as a corporal, because that rank progression is viewed as technical progression. When this delay was imposed on the infantry, it was completely unsatisfactory. It took considerable time to get it accepted that we must give accelerated promotion to privates. Canada has just won the annual NATO section competition - an R22eR section I mention in passing! That section leader was a fellow who has been in the Army for less than five years. He could not have been in that position if we had not finally decided in our unified force that in the combat arms rank is not the same thing as in the technical trades. In the technical trades rank is much more a function of technical skills than of the 'get up and go' and authority that you want in a battlefield leader.

MGen Robert Larose, Associate Director of the Association of Canadian Community Colleges

My concern is the trend in the Canadian army towards the non-employment of women in the reserves. In future, when our demographic problems increase, we are going to find ourselves compelled to call upon women more and more. What disturbs me is that in the militia right now we are downgrading the need for women. We are now facing high unemployment and we are taking advantage of that to say to women that they are no longer needed. What will happen in the longer term when we find that we need them, but have not developed and evolved their employment in all jobs which are not strictly combat? There are a lot of support jobs within battalions, regiments, brigades, that could be filled by women quite effectively.

What are the current policies concerning the employment of women in the reserves?

Liston

While I am not in a position to give you detailed information on our personnel policies on this issue, I would suggest that in the context of looking at the future there are innumerable opportunities for women to serve. In my mind, I divide the problem into two separate issues. The first is that of getting women into occupations in which they do not normally serve. That's more of a social issue than a manpower or womanpower planning issue. I don't think I'm in a position to address it. The other issue involves the question of whether we will require women and will there be a place for them in mobilized forces? Definitely. In the last war, women did all sorts of things, including manning ack-ack guns in Britain. There is no doubt that we will need them.

Talking as an observer, I used to get furious when I saw 19-year-old soldiers driving staff cars in Ottawa instead of being in a field unit. But the whole staff car fleet now appears to be run by women, including the supervisors. There would seem to be positive change taking place right now.

Jean Fournier, NDC Course #1

I joined the NPAM (the 'not particularly anxious militia' as it was known) in 1934. The question I have to ask is probably for someone of my age group – Colonel Liston is much too young!

A lot of concern has been expressed in our seminars about the state of readiness for war. The question I would like to ask is were we better prepared in

August 1939? My battery was mobilized at 0001, September 1. That means we had some semblance of a mobilization plan, I presume. Were we better prepared then than we are in November, 1982?

Liston

We are much better prepared now. The only problem is that the job is completely different. In 1939, with the exception of those who landed in Calais and then came right back, and those in the Dieppe raid, Canadians didn't go into battle until the landings in Sicily. (Actually there was Hong Kong as well, but I wouldn't want to give that as an example.)

The problem today is that we are not talking about mobilizing the Canadian forces to fight a war several years later. We are talking about troops on two hours' notice to move to the Czechoslovakian border; and we're talking about a requirement to replace their casualties within four or five days from troops in Canada. We are talking about the requirement to sustain those forces for as long as we wish to avoid nuclear war. How long can we afford to avoid bringing out the nuclear threat? I don't know. But that's the problem of comparing our position now to World War II.

Friedman

I think the two papers go together in a very interesting way. If you don't train your rear area people as troops, if they don't think of themselves as troops, and the war is as fluid as Air/Land 2000 makes it look – but it doesn't go our way at first – then those rear area troops are exposed to attack quite early. I keep thinking of the policy of the Marines: everybody, no matter what his specialty, was initially an infantryman. They found that very valuable in places like Guadalcanal where the cooks often did end up having to fight.

The next war is going to move very fast. The Russians are set up to move fast, and if we go to the 2000 concept, so are we. I would think a lot more rear area troops would end up fighting as troops at some point pretty early. I wonder if that has worried anybody?

Liston

I'm not suggesting that they should not have to fight as troops. I'm only suggesting that in the combat arms we must stick to our standards and we cannot allow them to be watered down in compromises to meet the requirements of the technical trades. I have not said that the technical tradesmen should not have a minimum standard of being able to protect themselves. There are, however, a great number of technical tradesmen who will not be in the rear area of the army. Indeed, I did suggest that the army logisticians and mechanics and so forth would have to have a much more demanding field orientation than perhaps technicians who would spend the war supporting static air bases for NORAD, for example. You must remember we have a unified force. When we talk about common training it includes the whole range of tracks from infantrymen to dental technicians or static bases in Canada. We are all one force.

Brian MacDonald

May I ask a supplementary on that one? If we have a unified force and we are training technicians, some of whom are to be employed with the army out in the long wet grass and others who will be employed in a totally different environment such as an air base, how do we then train that collection of technicians? Do we separate them into army technicians and air force technicians?

Liston

Personally, I'd rather avoid that issue! It's something that the technical trades are struggling with themselves. Should they have a green identity or an environmental identity?

Coming back to Dr Friedman's point, the attrition rates in the Marines, the last time I saw them, were about 42 per cent. They are an excellent organization; but they are small; they are very demanding; and they are not able to retain their people to the same extent as the air force. I would suggest that if you were to impose on the United States Air Force the same demanding standards of soldiering that you impose on the Marines, that you wouldn't be able to keep all those technicians that you need so desperately.

Fulton

This business of 'austere support' is something that we're trying to analyze. We found out that heavy divisions as currently organized contained logistics structures in the rear that turned out to be their own best customers. That's where most of the pile-up of troops was. We have to cut down this large logistics base, whether it be maintenance organization or hospitals, or whatever.

The only way we will get at that is through some sort of technological solution that reduces, for example, the high inventory of ammunition. We could reduce the ammunition load by using more 'smart' munitions with first hit capability, or by things like the electric gun that has no propellant charge. We are still using ammunition loaders too, a rather primitive way. There should be a way to package it, to modularize it, so that we can stick it right inside a tank. Why do we have to load a tank manually? That's a very good question that was not taken into consideration when we developed the M-1 but will be in the next systems. It will reduce the manpower bill and reduce some of the load.

High tech is an absolute necessity.

Beattie

I still think we are making too much of high tech against men. Gunners have lived with high tech throughout the ages. It is part of the artillery family. If you are going to be efficient in the field as a gunner you have to have a field artillery data computer, you probably have a laser range finder, and you are working with computers and sending data back to the guns. Gunners have been able to maintain their regimental esprit de corps and to be combat soldiers.

I think our people can measure up. And I don't think generals need to be protected against black boxes or vice versa. I think we are equal to the task, whatever it is, no matter how technical the equipment. If we get into a hot battle, it won't be very long before the high tech becomes low tech, as you run out of these fancy weapons systems. Then the abilities of fighting and soldiering are going to be paramount.

Liston

Your point is well taken, General Beattie. I guess the main thrust of both our presentations was that people will be using high tech and that they are quite capable of it. The problem of high tech is not with the user; it lies in reprogramming and repairing the computer. This is where new trades are being developed and brought into the forces.

Fulton

The principal problem is development. Once it gets to the field, it is easily handled. The tanker in the new M-1 tank is still a tanker – he's not a high tech tanker. I think that's the point you are making, and I completely agree with you.

Dr D.A. Charters

The emphasis placed on the use of women in combat units suggested that we would only have them in the rear areas. But I think the rear areas in the very fluid and possibly deep battlefield are going to see a lot of violent activity. Can you put a woman in an infantry battalion or any unit within 50 miles of the FEBA and say 'You're probably not going to be attacked'? In that rear area there could be airborne units landing or almost anything. If you are attached to an infantry battalion, you have to think like an infantryman, and not assume that simply because you are a clerk that you are not going to be attacked.

Christopher Donnelley of the Soviet Studies Research Group has just produced an article on the Soviet operational manoeuvre group, which he interprets as a new trend in Soviet manoeuvre tactics. The whole emphasis there is that they would like to get into the rear areas very quickly to completely disrupt the whole administrative structure. If you put women in there, they will have to think like combatants from the moment they step off the plane.

Liston

The only comment that I can make on that is that the problem will probably solve itself. The moving of women into non-traditional areas is basically one of trying to change social attitudes rather than meeting an operational need. I think that in practice, once the doors are opened, there are not many women who go through. We have great trouble in even carrying out our experiments with women in non-traditional roles, because so few are interested. For that reason, more than any doctrine or rationale about where women should or should not serve, I think it will solve itself in the long run.

Brian MacDonald

Colonel Fulton and Colonel Liston, the clearest indication of the quality of your presentations is the question period that has followed. We've had something like an hour of intensive questioning and comments, and I think you may take that as the accolade of having made two very successful presentations. Thank you very much.

I would now call upon Dr Bell to provide the summation for this seminar.

GEORGE BELL

SUMMATION

Ladies and gentlemen: We have indeed had two days of stimulation. In the education world we talk about mindpower these days. Here we have been in a period of mind stretch. We've had a smörgasbord of ideas.

In summing up this session, I will start by saying that we have had so much material delivered in terms of ideas that the subjects of a number of subsequent seminars are already evident. Our research group is going on immediately to a closed seminar with a number of others who will be dealing with directions for future research. No doubt they will take with them a number of ideas from this session.

Let us look at what we have heard - although I cannot cover everything. In Major Elliot's statement we heard about the difficulty of strategic deployment, the problems of projection of power. We had an indication of the vulnerability to interdiction of an area. If you look at the possibilities of interruptions in other areas that might affect the strategic materials required by Europe and North America to sustain themselves, you get some idea of the implications of conflict at strategic distance. From him and others we have heard of the increased impact of new weapons systems on the scale of casualties, and this leads to Colonel Liston's theme - the sustainability of forces.

A lesson for Canadians was expressed in the statement: 'We need to train for both opposed and unopposed landings.' Too many of our thoughts are based on reaching the area of operations without having to fight our way into it. We must consider how we get into an area: by sea mobility, by amphibious assault, or by airborne or helicopter landing.

Dr Charters and Michael Stevenson emphasized the aspect of low technology: the regimental system, the importance of the soldierly self, the battle skills, the combined arms tactics. Included too is the whole question of tactical force loadings - how you prepare yourself to go into operations. All of these things are evident needs, even in a high technology era.

Dr Friedman took us with both a sense of humour and a sense of cynicism through a critical analysis of the political aspects of defence capabilities. Defence capabilities are sometimes sacrificed by political decision - that probably has a familiar ring! He discussed the difficulties of a country like Britain in dealing with the Trident submarine question, because it is not necessarily a military decision related to an enemy. It might be a military decision related to Britain's view and France's view of themselves in relation to Europe and in relation to the United States.

Dr Friedman stressed too the need for a mobilization base. While he was talking about sea-lift capability, the ability to project power around the world by having a merchant fleet to call upon and having the automatic resupply capabilities of afloat navies, he also clearly reminded us that a mobilization base is necessary for the effective conduct of military operations. We spoke of external and internal lines, and I would like to remind you that we are always on external lines in Canada and the United States when we try to approach theatres of operations other than those in our own territories. Even in our own territory, the distances are so great that operations in the Arctic rim area are equivalent to crossing the ocean. We must think about what we need in the way of mobilization capability and in the capability to move ourselves strategically.

We have continuously emphasized that our forces lack low-level air defence. That deficiency was emphasized by Dr Friedman, by Major Elliot, and by Dr Lindsey, who stressed the need for rapid-fire, complex air defence systems. And we also need the capabilities of intelligence and detection, in order to tell where the enemy is and to be able to strike him and also to protect ourselves from his attacks. Early warning and electronic warfare become part of the normal mechanism of an armed force regardless of its size.

Just recently I argued in the Senate committee on maritime defence against Michael McGuire from Brookings Institute on the question of the CAST combat group operation in Norway and whether we should pull out of the Central Front and make Norway the area of total Canadian commitment - not because the idea was wrong, but because there are a number of political implications that do not allow it to be fulfilled. His statements, however, and the statements of Dr Friedman about the critical air positions in terms of the eastern Atlantic, should remind us clearly that there *is* a threat from Backfire bombers. There is an important mission, therefore, in holding Norway so that there will not be a land-based air capability that could strike even farther out at sea. There is a need to have air protection for the movement of that great part of the NATO commitment, which in the case of the United States is something like six divisions of the rapid reinforcement force. The strategic movement of our forces and our capabilities to fulfill our missions become important as part of that reinforcement, and so the high technology aspects (even the medium and low technology aspects) of those operations have to be recognized.

In the area of equipment development, George Lindsey took us on a tour d'horizon and in his usual lucid way set out a whole range of things that we must consider, some of which will be carried out by our strategic allies and others by us. He talked about the vulnerability of our bases (as did Colonel Liston this morning), of our air-heads and our bridge-heads, if we do not provide those technical capabilities for their defence against modern systems.

I wonder if we have been radical enough when talking about manpower. We still lack a national manpower policy. We lack an adequate mobilization plan and organization. We do not have a national service policy. We really do not have ready reserves. These concerns are not new. They have been expressed here before and in other circles and in articles written by John Hasek and others. We have to face up to the question of whether, in the manpower sense, we are able to meet the requirements of the rapid reinforcement concept. We have to be concerned about the *survivability* of the forces we have on the ground now, because they are undermanned and underdefended, and they do not have the necessary equipment to survive. We have to worry about their *sustainability*. Colonel Liston pointed out that the numbers required to maintain the force in battle for 30 days are well beyond the numbers Canadian governments have talked about in the past. I was pleased that, even with the limitations of the system, the numbers required are beginning to be identified publicly.

We have to deal with the security of bases and forces. The technicians, the people we have now in Europe on the bases in the air group, will have to defend their bases when the brigade goes to its emergency location. They do not have an air field defence group. The general defence of the rear areas, which is dependent upon the German southern army and reserves, is a *general* defence. They are unlikely to provide defence for our bases. We will have to do it.

So the question of readiness must be addressed. Readiness is a condition. We are not ready in a manpower sense. We are not ready in an equipment sense because the reserves don't have the equipment. Our war reserves are inadequate for sustaining our forces in operations in conventional warfare in the European theatre or elsewhere.

Our perception of the problem is wrong. We talk about our brigade group in Germany as being a *strategic* reserve. In fact, it is as close to the battlefield area as any other in-depth brigade within either the German or the American corps. Therefore it must be able to fight as a fully manned combat-capable unit. Not only is the enemy going to strike in its area, but the enemy is also going to strike the rear well behind it. Its readiness level and ability to actually move in operations in order to survive is a question we should look at. Our perception of our roles and how they are to be manned should be clarified.

In terms of technological developments, in the presentations here a whole series of ideas have been put to us. But let me remind you of prior options that we as Canadians have had. Mr Halton spoke about surveillance devices - but those ideas were around in 1957. That tilt-wing CL-84 has been around a long time. Does anybody remember the Avro car - the thing that looked like a flying saucer? Do you remember when the neutron weapon for the tank was talked about in Quebec City as early as 1957? We can write down lists of technological capabilities, but it brings up the question of the national political will to move ahead in the technological area and put the commitment behind it. Secondly, we have to deal with what has been an inhibiting foreign policy. In the whole field of military technological development and weapons exports, we have some national inhibitions which affect our abilities.

Gordon MacNabb last night stretched our minds on the question of the availability of human resources. This morning I looked at the NATO booklet on combinations of forces and comparative forces. That stressed that the technological lead we have had in the West over the Soviet bloc is fast closing. In the Soviet Union, science and technology doctoral degrees at the universities comprise about 80 per cent of university production. In the United States, they are 40 per cent of the university production of doctorates and in Canada they are much less.

When we look at the manpower data presented today, we should not be sanguine about the number of people going into community colleges or universities. If you look at enrolments in community colleges in terms of hard science, hard technology, you find the percentage is low. The numbers are increasing, but we have done nothing in terms of a national manpower policy or even a provincial manpower policy to indicate effectively the levels of shortage. As a university administrator, I have no document from any level of government that really prescribes need. Nor has industry been very good at indicating its long-term requirements. Right now, though our mining and metallurgical industries are suffering very badly, a problem for the future is developing. We are not producing enough people in those areas for the future, yet we know that the mineral processing capabilities of North America are going to be required for high tech industry.

We have discontinuities and disutilities in our thinking, because we are moved too often by short-term thinking.

We lack a strategic deployment capability in Canada. We have the capability to do more prepositioning, if we can get rid of the dollar curtain. We can do more in the area of strategic airlift. We can use that air industry that Mr Halton talked about if we will, in fact, commit the civil fleet, and also subsidize the civil fleet to have within it the cargo carriers that we will need for strategic deployment.

Colonel Liston referred to a recently published article by General Rogers, the Supreme Allied Commander Europe, which is relevant to the impact of technology on the battlefield. In that presentation, he made a couple of telling statements. I will quote briefly from it because I think it is important in the context of this seminar.

We must not delude ourselves. NATO's continuing failure to fulfill its conventional needs means that we must now depend upon the use of theatre nuclear weapons to accomplish our mission of deterrence and defence.

There is a more acceptable alternative to this posture, namely to acquire conventional capability that would provide a good prospect of success in the forward defence of Europe. To be capable of such a defence does not demand that we match the Warsaw Pact land and air power on the continent division for division, tank by tank, or plane by plane, nor does it entail a buildup of such proportions that it might be construed as an offensive threat to the Warsaw Pact. We do need some increase in the number of forces to offset Soviet military growth, but far more important to success is the enhancement of our ability to do better with our forces in being and to carry out the essential modernization of these forces.

In operational terms, we need the capability to hold the lead divisions of a major Warsaw Pact conventional attack while we conduct an effective interdiction and destruction campaign with conventional means against his follow-on forces, thereby neutralizing them before their weight can be brought to bear at the front lines. Such capability will also provide us the opportunity to carry out counterattacks to reverse the flow of battle. We must also exploit electronic warfare to a much greater extent than we do now, so that we can blind our adversary's radar and interrupt the central direction of his operational forces. At the same time, we must have greater assurance in our ability to control the sea lanes to Europe and to win the air war over the continent.

The concept for conventional destruction of the Pact follow-on forces has been developed, utilizing the good work in this field of several of the NATO nations. The concept is now part of the ACE doctrine. Our current challenge would be to exploit it as best we might with present conventional means. Meanwhile major efforts must be expended to encourage nations to develop and procure the advanced technological targeting and weapons systems that will make the application of the concept most effective. The provision of such a robust conventional capability for NATO not only would enhance deterrence and raise the nuclear threshold, but would face the Warsaw Pact, should it attack conventionally and its attack be frustrated, with either having to be first to escalate to theatre nuclear weapons or withdrawing their forces.

As Canadians we often (as General Kirby so aptly pointed out) look at the scene that Colonel Fulton put in front of us through the wrong end of the telescope. I ask you to look at the country, look at the world, both for the future and for the past. A question would be: Is small beautiful? We get patted on the head in NATO for having the most professional, highly trained forces, which continues to be illustrated in competition, as suggested by Colonel Liston. But – here comes the but – we need more of them.

I suggest to you that we must also look at the question politically – not just at the Soviet bloc in Europe, but also at Third World forces. Some Third World countries now have an increasing technological capability. Some of the countries in Latin America – Brazil for example – can produce in five areas of weapons production. Others can produce in three or four. There is now a capability of selling weapons back and forth in the Third World. There is now, as well, a capability of conflict in the Third World. There is the possibility of Third World countries being far more capable than we are because we have not kept up with the state of the art and the pace of development.

Can we responsibly expect that in a major emergency a country like Canada will get away with such a small commitment? I think we should plan *at least* at the corps level to ensure that we have the capabilities in the forces we do deploy, to ensure that we buy or share with an ally a capability such as we are doing with AWACS, or that we develop capabilities on our own, so that forces that we deploy are complete. In looking at our reserves, at business and industry, we must see what frameworks in the high tech areas can be embodied within corporations as shadow reserve units that can be brought forward in mobilization.

We need to deal with all this with greater national will, greater national leadership. We need to deal with facts in debate. A debate has just taken place concerning nuclear disarmament, but with all the high technology we have, we have not explained to the public, who have been asked to vote on those referenda, the real questions that must be answered. We have left a vacuum and we need to fill it.

We need, in our planning, to remove the dollar curtain. When we are talking about high tech, we are talking about the potential in war. If we go to war, we will employ the economic energy and resources and the human material of the nation. We will find it possible to shift our priorities. But we are missing opportunities in peacetime to reinforce our forces in terms of technical training. We are missing opportunities to expand and use the armed forces training base to effectively improve the national capability. We have not taken advantage of our unemployment situation to deploy any dollars to increase the reserves, let alone the regular forces. One could go on with a number of lessons, but I will stop here.

I would like to express our thanks to Major Elliot for the successful national tour which he made on our behalf, and for his erudite presentation here. I would like to thank Dr Charters and Dr Friedman for taking our minds into other areas of the Falklands situation to help us focus on the issues that confront us for the future. I only wish we could have expanded our discussion to include something on the Iran-Iraq war and the operations in Lebanon. I want to thank George Lindsey for his constant support of the Institute, for his willingness to express his very extensive knowledge and lucid mind to us, not just today but on many occasions in the past. I thank Mr Halton for his excellent presentation on one of our key industrial sectors and its ability to contribute to our future development. I express our thanks to Dr MacNabb for his talk on human resources, and to Colonel Liston for his thoughtful and indeed courageous paper on military manpower - and I can assure him that one of our seminars in the not too distant future will be on the subject of military manpower and will follow on from the manpower seminar we had before. I want to thank Colonel Fulton for bringing us that very clear, precise presentation of the thinking that is going on in NATO. He also used the term 'ABC' - America, Britain and Canada - the closest grouping in the area of combat development that has existed since right after World War II. This is affecting the combat development thinking and the development of the corps concept in Canada. I thank him very much for that presentation which reminded us that we are not just locked into Europe in the future. There are other continents and other climates to prepare for, and we have to provide for both high and low intensity war.

I would like to recommend to you the excellent article by General Rogers which appeared in *Foreign Affairs* in the summer of 1982, entitled 'The Atlantic Alliance: Prescriptions for a Difficult Decade.' There are also two very short articles in the November issue of *The Reader's Digest* on pages 49 to 54, one dealing with perspectives on the peace movement and the other on deterrence versus the freeze. They are short, succinct and thought-provoking. If you want to dip a little deeper into the morass of how to think or not to think about nuclear war, there is an excellent article in *Commentary* by Edward Luttwak, 'How to Think about Nuclear War,' which presents some of the arguments that underlie the current exchange of ideas between East and West.

The Spring Seminar on May 6 will deal with Canada's strategies for space. It will carry on from the high tech area and will deal with aspects of satellites, lasers, and sensors in space. We will look at the impact of the space dimension on Canadian national security in its broadest sense, on Canadian defence policy and on Canadian industrial strategy.

I thank you for coming. I thank all the speakers for making this such a stimulating seminar. I know that many of the subjects raised will be the ideas for discussion in the future.

I declare the seminar closed.

GLOSSARY

ABCA	America-Britain-Canada-Australia
ACE	Allied Command Europe
ADM	Assistant Deputy Minister
AMTI	Airborne Moving Target Indication
APC	Armoured Personnel Carrier
ASW	Anti-Submarine Warfare
ATGM	Anti-Tank Guided Missiles
AWACS	Airborne Warning and Control System
CAST	Combined Air Sea Transportable
CDS	Chief of the Defence Staff
COTC	Canadian Officers Training Corps
CW	Chemical Warfare
DDH	Helicopter-equipped Destroyer
DND	Department of National Defence
EW	Electronic Warfare
FCO	Foreign and Commonwealth Office
FOO	Forward Observation Officer
GIUK	Greenland-Iceland-United Kingdom
LCU	Landing Craft Utility
LOTRP	Land Operations Trade Reassignment Programme
MITCP	Militia Individual Training and Career Profile
NIAG	NATO Industrial Advisory Group
NORAD	North American Air Defence
NSERC	Natural Sciences and Engineering Research Council of Canada
PVC	Polyvinyl Chloride
RDF	Rapid Deployment Force
Ro-Ro	Roll On-Roll Off
RPV	Remotely Piloted Vehicle
SAM	Surface to Air Missile
SLAR	Sideways Looking Radar
SCEAND	Standing Committee on External Affairs and National Defence
VSTOL	Vertical and Short Takeoff and Landing
VTOL	Vertical Takeoff and Landing